Lecture Notes on Multidisciplinary Industrial Engineering

Series Editor

J. Paulo Davim, Department of Mechanical Engineering, University of Aveiro, Aveiro, Portugal

"Lecture Notes on Multidisciplinary Industrial Engineering" publishes special volumes of conferences, workshops and symposia in interdisciplinary topics of interest. Disciplines such as materials science, nanosciences, sustainability science, management sciences, computational sciences, mechanical engineering, industrial engineering, manufacturing, mechatronics, electrical engineering, environmental and civil engineering, chemical engineering, systems engineering and biomedical engineering are covered. Selected and peer-reviewed papers from events in these fields can be considered for publication in this series.

Razvan Tamas · Cristina Adelaida Heiman
Editors

Selected Papers from AToMS 2024

Applications in Environment Engineering, Material Science, Mechanical Engineering, and Social Impact of the Technology

Springer

Editors
Razvan Tamas
Department of Electronics
and Telecommunications
Constanta Maritime University
Constanta, Romania

Cristina Adelaida Heiman
Department of Electronics
and Telecommunications
Constanta Maritime University
Constanta, Romania

ISSN 2522-5022 ISSN 2522-5030 (electronic)
Lecture Notes on Multidisciplinary Industrial Engineering
ISBN 978-3-032-02451-0 ISBN 978-3-032-02452-7 (eBook)
https://doi.org/10.1007/978-3-032-02452-7

This Springer imprint is published by the registered company Springer Nature Switzerland AG
The registered company address is: Gewerbestrasse 11, 6330 Cham, Switzerland

Contents

Part I
Applications in Environment Engineering

Chapter 1
Background Radiation Monitoring

Sabina Nedkova, Plamena Atanasova, Stela Naydenova, Mariana Panaitescu, and Fanel-Viorel Panaitescu

Abstract Radiation risk cannot be limited by territorial boundaries. This fact and the long exposure time of radioactive elements (most of which have a decay period of tens to hundred thousand of years), require measures of timely assessment of the level of risk, with the aim of prevention and control, called radiation monitoring. The study is based on monitoring of the gamma radiation background and studying of its relation with the surface contamination over three monitoring points and the surface contamination, detected on the air filters of $PM_{2.5}$ and PM_{10}. The research is performed with personal portable dose rate meter Radiagem 2000 and Alpha/Beta/Gamma external CSP probe at two ecosystems—terrestrial and marine within the Black Sea coastal zone area. There is a positive correlation between the amount of the particles $PM_{2.5}$, deposited on the filters and the surface contamination, which can be related with the presence of the black carbon, acting as an absorbent.

Keywords Assessment · Black carbon · Energy · Monitoring · PM2.5 · PM10 · Radiation · Soil · Source · Risk

1.1 Introduction

The radiation background creates the environment, which living organisms inhabit and which affects them, from the time Earth had been formed as a solid planet, more than 4.5 billion years ago. Within the time since then, the influence of the radiation

S. Nedkova (✉) · P. Atanasova
Department of Technologies, Materials and Material Science, Technological Faculty, Burgas State University, Burgas, Bulgaria
e-mail: sabina_nedkova@abv.bg

S. Naydenova
Department of Environmental Protection, Faculty of Natural Science, Burgas State University, Burgas, Bulgaria

M. Panaitescu · F.-V. Panaitescu
Faculty of Naval Electromechanics, Constanta Maritime University, Constanta, Romania

R. Tamas and C. A. Heiman (eds.), *Selected Papers from AToMS 2024*,
Lecture Notes on Multidisciplinary Industrial Engineering,
https://doi.org/10.1007/978-3-032-02452-7_1

background has been a factor in the development of living organisms, having in its low-dose's values a distinct effect on the hormesis and their adaptive response to the factors of the environment [1]. As a factor of high importance, the ionizing radiation must be subject of a continuous monitoring in order to evaluate and control its effect on the life and health of living organisms. All radionuclides, which form the ionizing radiation of the environment, can be divided into four main categories: Nature born—Primary, created during the formation of the Earth; Radiogenic—Progeny of primary radionuclides; Cosmogenic—obtained as a result of the interaction of cosmic rays with terrestrial chemicals; and Anthropogenic, resulting from human activities [2, 3].

A. Gamma radiation background

The presence of natural and anthropogenic radionuclides in the atmospheric air serves as an indicator for determining of atmospheric radioactivity, or "gamma background/γ- background", as an indicator that describes the impact of the environment on living organisms, namely: "power of the equivalent dose of gamma radiation from the air". The natural γ-background is a physical feature of the environment specific to a given area. Above the earth's surface the aerial γ-background is created mainly by the radionuclides contained in the upper soil layer about 30 cm thick. For this reason, an essential role in the formation of the γ-background is played by the natural radionuclides of terrestrial origin from the three-natural radioactive "families": the uranium-radium family with the progenitor ^{238}U, the thorium family with the progenitor ^{232}Th and the actinium family with the progenitor ^{235}U, which include a total of 47 different radionuclides and have a wide distribution in the earth's crust [2]. The gamma background of the environment is measured in micro Sieverts per hour [$\mu Sv.h^{-1}$].

B. Surface contamination

Radioactive substances could enter the biosphere as a result of testing of nuclear devices in various natural environments, or in the form of radioactive waste from industrial, power reactors, or from uncontrolled emissions of radionuclides into the external environment as a result of accidents in a nuclear environment. A limited amount of radioactive materials enters the environment as a result of an extraction of uranium ore. Radioactive volatile substances spread in a form of a radioactive cloud, consisting of particles of various sizes, deposited in the form of radioactive sediments, the distribution of which depends mainly on weather conditions [4].

1.2 Material and Methods

The subject of the present research is a radiological monitoring, which was developed as:

- a monitoring of a surface contamination on samples of air particular matter's fractions $PM_{2.5}$ and PM_{10}, from air analysis of an urban background point in [$\mu g.m^{-1}$];

- a monitoring of gamma radiation background and ambient dose equivalent rate, measured with personal portable dose rate meter in [μSv.h^{-1}];
- a monitoring of a surface contamination over land and sand in [Bqeq.cm^2], from two monitoring locations in terrestrial and marine environmental systems;

The monitoring locations for gamma radiation background and surface contamination are selected to represent two ecosystems—a terrestrial (a monitoring point within University Prof. d-r Asen Zlatarov's campus) and a marine (two monitoring points at the Burgas beach) environmental system, in the town of Burgas, Bulgaria. The collected data includes gamma radiation background/air doze rates, registered on the three monitoring points and surface contamination, over dry and wet sand in the monitoring location on the beach and over a land on the monitoring location within the university for a period of six months. The gamma radiation background is measured as H*(10)—ambient dose equivalent rate in μSv.h^{-1} with a personal portable dose rate and survey meter Radiagem 2000. H*(10) reflects the absorption of gamma radiation in a tissue-equivalent area (equivalent to the human body), i.e. through H*(10), the human absorbed dose can be directly calculated.

The monitoring of the surface contamination or the presence of a radioactive substance on a land and sand surface, is measured by portable survey meter Radiagem 2000 and Alpha/Beta/Gamma external 15 cm^2 CSP probe w/pancake G-M NOM 0,006,364 SABG-15+. Contamination in this case is explained as the presence of a radioactive substance on a surface in quantities in excess of 0.4 Bq.cm^{-2} for beta and gamma emitters and low toxicity alpha emitters, or 0.04 Bq.cm^{-2} for all other alpha emitters [5].

The monitoring of a surface contamination for air particular matter's fractions $PM_{2.5}$ and PM_{10} is measured in µg.m^{-3}on filters from an urban background monitoring point, distant from specific sources of pollution but quite near the land's monitoring point for gamma radiation and surface contamination within the university's campus. The matter is collected at the period 22nd of January to 18th of February 2024. The PM samplers are mounted approximately 10 m above the ground. The $PM_{2.5}$ samples were collected through DadoLab Giano BC1 equipment, and a sequential PM sampler with Black Carbon analyzer in real time. The PM_{10} samples were collected through DadoLab Giano and Gemini. Both devices collected PM on WhatmanVR QM– quartz filters, 47 mm for 24 h with flow rate about 2.3 Nm^3.h^{-1} [6–8]. All collected samples were stored at aluminum foil until surface pollution's analysis. Fine particulate matter's fractions ($PM_{2.5}$ and PM_{10}) are considered the most dangerous air pollutant in Europe due to their toxic effects, stemming from both its size and composition. In the present research is studied the relation between the surface radioactive contamination on an air filter, the presence of the so called "black carbon" in it, the PM size, the amount of fine particulate matter's fractions $PM_{2.5}$ and PM_{10} and the gamma background. The data was processed and analyzed by the SPSS statistical data processing tool.

Table 1.1 Correlation amount of $PM_{2.5}$, surface contamination, gamma background

	Amount of $PM_{2.5}$ [$\mu g.m^{-3}$] (1)	Surface contamination [Bq $eq.cm^{-2}$] (2)	Gamma background [$\mu Sv.h^{-1}$] (3)
(1)	1	0.393*	0.173
(2)	0.393	1	0.333
(3)	0.173	0.333	1

* $p<0.05$

1.3 Results and Discussion

The reliability of the data gathered had been analyzed and reported with Cronbach's alpha coefficient of moderate to good value of 0.697. This indicates good internal consistency among the items: surface contamination, gamma background, and the amount of the PM. The mean value of gamma radiation is within the safe limits (MV $= 0.089\ \mu Sv.h^{-1}$).

1.3.1 Surface Contamination of $PM_{2.5}$. Correlation with the Amount of $PM_{2.5}$ and Gamma Radiation Background

There is a positive correlation among the three data types-the amount of the particles $PM_{2.5}$, deposited on the filters, the surface contamination over the filters and the gamma background (Table 1.1).

1.3.2 Surface Contamination of PM10. Correlation with the Amount of PM 10 and Gamma Radiation Background

There is a negative correlation among the three data types-the amount of the particles PM_{10}, deposited on the filters, the surface contamination of the filters and the gamma background (Table 1.2).

Table 1.2 Correlation amount of PM_{10}, surface contamination, gamma background

	Amount of PM_{10} [μg.m^{-3}] (1)	Surface contamination [Bq eq.cm^{-2}] (2)	Gamma background [μSv.h^{-1}] (3)
(1)	1	−0.123	−0.014
(2)	−0.123	1	0.251
(3)	−0.014	0.251	1

1.3.3 Analysis on the Surface Contamination of the Filters of $PM_{2,5}$ and PM_{10}, and Their Relation with the Amount of $PM_{2,5}$ and PM_{10} and Gamma Radiation Background

The analyzed data include two types of filters for $PM_{2.5}$ and PM_{10} dust particular matter. The positive correlation results are detected on the $PM_{2.5}$ filters, which contain Black Carbon (BC) includes in the chemical composition of $PM_{2.5}$. The positive relation at that case with the surface contamination, or the presence of a radioactive substance, could be explained, with the physical qualities of the black carbon, which could work as a carrier or adsorbent for radioactive particles, allowing them to adhere to the surface of the particle and from there to the filter more easily. This could result in a hypothetical accumulation of radioactive contaminants on the filter, increasing the overall surface contamination.

The surface contamination could be of alfa or beta type. The authorized level of beta contamination for non-radiologically controlled area is normally higher, than for alpha contamination. The device, used for the research (Alpha/Beta/Gamma external 15 cm^2 CSP probe), could not specify the type, but as seen (Table 1.3), the mean values of the detected surface contamination of both $PM_{2.5}$ and PM_{10} dust filters are close to the limit of beta type of contamination 0.4 Bq.cm^{-2}. At that case is obvious, that the units of measurement Bq.cm^{-2} and Bq eq.cm^{-2} are different. Bq.cm^{-2}, represents the activity of a radioactive source in Becquerel (Bq) per unit area (cm^2) and Bq equivalent Bq eq.cm^{-2} refers to the dose equivalent of radiation emitted by a radioactive source, depending on the type of radiation being emitted and its biological impact on human health and the potential health risks associated with exposure to radiation. Since Bq eq.cm^{-2} takes into account a broader range of factors than Bq.cm^{-2}, it can be considered greater than or equal to Bq.cm^{-2}.

It is seen (Table 1.3) that $PM_{2.5}$ filters, which contain BC, deposited on them had been assessed with higher mean value, than the PM_{10}filters, with no BC, which supports the thesis of a possible role of the black carbon as adsorbent for radioactive particles, allowing them to adhere to the surface of the dust filter more easily. BC is formed through the incomplete combustion and has significant effects on climate and human health [9, 10]. BC consists of pure carbon in several forms, but the relevant particle size fraction can include carcinogens and other toxic species as well [11].

Table 1.3 MV and SD of PM2.5 and $PM_{10,}$ surface contamination

		Surface cont. PM $_{2.5}$ [Bq eq.cm^{-2}]	Surface cont. PM_{10} [Bq eq.cm^{-2}]
N	Valid	28	28
	Missing	0	0
Mean		0.46	0.40
Mode		0.44	0.37
St. deviation		0.07	0.06

1.3.4 Analysis of the Gamma Background and the Surface Contamination of Sand

The data on gamma radiation and surface contamination of a marine environmental system had been collected within six months, from January to June 2024, on two accessible monitoring points on wet and dry sand. The data includes monitoring of gamma radiation background and ambient dose equivalent rate, measured with Radiagem portable dose rate meter in μSv.h^{-1} and surface contamination over dry and wet sand in Bq eq.cm^{-2}, measured with external alfa, beta, gamma probe (Table 1.4). The analyses (Table 1.5), shows that there is a positive correlation between the gamma radiation background and the surface contamination, detected over a dry sand's monitoring point, and a negative correlation between the gamma radiation background and the surface contamination over a wet sand's monitoring point. The correlation analysis shows high statistical significance of the relation between the gamma background and the dry sand's contamination.

Table 1.4 Gamma radiation and surface contamination over dry and wet sand

Months	Gamma radiation [μSv.h^{-1}]	Surface contamination dry sand [Bq eq.cm^{-2}]	Surface contamination wet sand [Bq eq.cm^{-2}]
January	0.09	0.38	0.28
February	0.07	0.37	0.25
March	0.10	0.39	0.29
April	0.11	0.39	0.31
May	0.07	0.18	0.37
June	0.07	0.28	0.38

Table 1.5 Correlation gamma background, surface contamination on dry and wet sand

	Gamma radiation [μSv.h^{-1}] (1)	Surface contamination dry sand [Bq eq.cm^{-2}] (2)	Surface contamination wet sand [Bq eq.cm^{-2}] (3)
(1)	1	0.673	−0.330
(2)	0.673	1	−0.802
(3)	−0.330	−0.802	1

1.3.5 *Analysis of the Gamma Background and the Surface Contamination of Land*

The data on gamma radiation and surface contamination of terrestrial environmental system had been collected within six months, from January to June 2024, on accessible points within monitoring location at the University Asen Zlatarov's campus.

The radiological monitoring, includes monitoring of gamma radiation background and ambient dose equivalent rate, measured with Radiagem portable dose rate meter in μSv.h^{-1} and surface contamination over land in Bq eq.cm^{-2}, measured with external alfa, beta, gamma probe (Table 1.6). The analysis (Table 1.7) shows that there is a positive correlation between the gamma radiation background and the surface contamination, detected over a land's monitoring point. The correlation analysis shows higher statistical significance of the relation between the gamma background and surface contamination over dry sand, than the one between the gamma background and the land's contamination.

Table 1.6 Gamma radiation and surface contamination over land

Months	Gamma radiation [μSv.h^{-1}]	Surface contamination [Bq eq.cm^{-2}]
January	0.08	0.34
February	0.07	0.30
March	0.09	0.35
April	0.10	0.35
May	0.11	0.37
June	0.06	0.40

Table 1.7 Correlation gamma background, surface contamination over a land

	Gamma back ground [μSv.h^{-1}] (1)	Surface contamination over land [Bq eq.cm^{-2}] (2)
(1)	1	0.016
(2)	0.016	1

Table 1.8 Mean and mode values on Gamma radiation background at the beach, Surface contamination of dry sand, Surface contamination of wet sand, Gamma radiation background over a land and Surface contamination of land for a period of six months

N	Valid	Gamma backround beach	Surface contamination dry sand	Surface contamination wet sand	Gamma background land	Surface contamination land
		6	6	6	6	6
	Missing	0	0	0	0	0
Mean		0.09	0.33	0.31	0.09	0.35
Mode		0.07	0.39	0.25	0.06	0.35

The statistical analysis (Table 1.8), on the mode and mean values of the measured data—Gamma radiation background at the beach, Surface contamination of dry sand, Surface contamination of wet sand, Gamma radiation background over a land and Surface contamination of land, shows, that mode values are within the safe limits of both gamma radiation background (0.06–0.40 $\mu Sv.h^{-1}$) and surface contamination (0.4 $Bq.cm^{-2}$ for beta and gamma emitters and low toxicity alpha emitters).

1.4 Conclusions

The analysis of the data from the radiological monitoring developed on the two monitoring locations in marine and terrestrial ecosystems (marine location –dry and wet sand's monitoring points and terrestrial location- land point) as:

- surface contamination on samples of air particular matter's fractions $PM_{2.5}$ and PM_{10}, from air analysis of an urban background point;
- a monitoring of gamma radiation background and ambient dose equivalent rate;
- Surface contamination over wet and dry sand and over a land.

Brought us to the following results:

- The black carbon could have a possible role as an adsorbent for radioactive particles, allowing them to adhere to the surface of the dust filter of $PM_{2.5}$ more easily and causing higher surface contamination;
- There is a positive correlation between the gamma radiation background and the surface contamination, detected over dry sand's monitoring point, and a negative correlation between the gamma radiation background and the surface contamination over wet sand's monitoring point. The correlation analysis shows high statistical significance of the relation between the gamma background and the dry sand's contamination;
- There is a positive correlation between the gamma radiation background and the surface contamination, detected over the land's monitoring point. The correlation

analysis shows higher statistical significance of the relation between the gamma background and surface contamination over dry sand, than over land.

- The mean values of the detected surface contamination of both $PM_{2.5}$ and PM_{10} dust filters are close to the limit of beta and gamma emitters and low toxicity alpha emitters—0.40–0.46 [Bq eq.cm^{-2}] and could be considered as safe.
- The gamma radiation background, measured as H*(10)—ambient dose equivalent rate in μSv/h over the sand and the land is within the safe limits of 0.06 to 0.40 [μSv/h].
- The surface contamination detected over wet and dry sand and land is less than, the limit of beta and gamma emitters and low toxicity alpha emitters type of contamination—0.31–0.35 [Bq eq.cm^{-2}] and could be considered as safe.

Acknowledgements The present research was supported by the project BLOW-Black Sea Floating Offshore Wind, Grant Agreement **N°: 101084323**

References

1. Jin, S., Jiang, H., Cai, L.: New understanding of the low-dose radiation-induced hormesis. Radiat. Med. Protect. **1**(1), 2–6 (2020). ISSN 2666–5557
2. Borisov, G. et al.: Medicine of the Catastrophes. Knowledge publishing house, Study book (1995)
3. Angelov, Hristo Assoc. Prof. Dr. Head of the research team: Study of the combined impact of natural radioactive background Research project funded by the Research Fund under Contract DN 04/1 of 13.12.2016, BEO Musala INRNE-BAS, (2017–2018)
4. Bajenov, V.A. et al., Vrednje himicheskie veshtestva, Radiosktivnie veshtestva, Leningrsd, Chemistry (1990)
5. Regulations for the Safe Transport of Radioactive Materia 2018 Edition, Specific Safety Requirements No. SSR-6 (Rev. 1)
6. Naydenova, S., et al.: Atmospheric levels, distribution, sources, correlation with meteorological parameters and other pollutants and health risk of PAHs bound in PM2.5 and PM10 in Burgas, Bulgaria–a case study. J. Environ. Sci. (2022)
7. Hristova, E., et al.: PAHs and Black Carbon in Urban Air Particulate Matter in Bulgaria, Environmental Protection and Disaster Risks. Springer International Publishing
8. McDuffie, E.E., Martin, R.V., Spadaro, J.V., et al.: Source Sector and Fuel Contributions to Ambient PM2.5 and Attributable Mortality Across Multiple Spatial Scales. Nat Commun **12**, 3594 (2021)
9. Bond, T.C., Doherty, S.J., Fahey, D.W., Forster, P.M., et al.: Bounding the Role of Black Carbon in the Climate System: A Scientific Assessment. JGA Atmos. **118**(11), 5380–5552 (2013)
10. Gerlos-Nijland, M., Lanki, T., Salonen, R., Cassee, F.: Health Effects of Black Carbon. Bonn: WHO Regional Office for Europe, The WHO European Centre for Environment and Health (2012)
11. Bessagnet, B., Allemand, N.: Review on Black Carbon (BC) and Polycyclic Aromatic Hydrocarbons (PAHs) Emission Reductions Induced by PM Emission Abatement Techniques. TFTEI Techno-Scientific Secretariat, Paris, France, p 156 (2020)

Chapter 2
Experimental Research Regarding the Hydrocarbons in the Romanian Black Sea Waters Area

Florentina A. Vasilica, Carmen A. Safta, Mariana Panaitescu, and Fănel V. Panaitescu

Abstract Experimental research on the analysis of hydrocarbons in water in the Black Sea area refers to the study of the impact of hydrocarbon pollution on the marine environment and biodiversity. Hydrocarbons are organic substances made of carbon and hydrogen atoms, which can come from natural or anthropogenic sources. Hydrocarbon pollution affects the quality of water, sediments and marine organisms, with negative consequences for human health and the economy. Analyses of hydrocarbons in water in the Black Sea are aimed at assessing the degree and extent of pollution, identifying sources and types of hydrocarbons, to monitor their evolution and degradation over time, to estimate risks and effects on the environment and human, to propose measures to prevent, control and remedy pollution. For the 2023 period, we took specific water samples from nearby areas, which were sent for testing to the accredited laboratory for PAH determinations in water. For the period 2021–2023, we analyzed and compared the data received and those obtained with the values imposed by law, thus obtaining the chart for their standard deviation.

Keywords Analysis · Hydrocarbon · Polycyclic aromatic hydrocarbon · Black sea

2.1 Introduction

Experimental research on the analysis of hydrocarbons in the Black Sea waters area refers to the study of the impact of hydrocarbon pollution on the marine environment and biodiversity. Hydrocarbons are organic substances made of carbon and hydrogen

F. A. Vasilica (✉) · C. A. Safta
National University of Science and Technology POLITEHNICA Bucharest, Bucharest-Romania, Bucharest, Romania
e-mail: vasilica.florentina16@gmail.com

M. Panaitescu · F. V. Panaitescu
Faculty of Naval Electromechanics Engineering and Environmental Protection in Industry, Constanta Maritime University, Constanta, Romania

R. Tamas and C. A. Heiman (eds.), *Selected Papers from AToMS 2024*, Lecture Notes on Multidisciplinary Industrial Engineering,
https://doi.org/10.1007/978-3-032-02452-7_2

atoms, which can come from natural or anthropogenic sources. Natural sources of hydrocarbons include oil and natural gas leaks from the seabed.

Anthropogenic sources of hydrocarbons include exploration, transport and processing of oil and natural gas, as well as oil accidents, accidental or intentional discharges, industrial and domestic wastewater [1].

Hydrocarbon pollution affects the quality of water, sediments and marine organisms, with negative consequences for human health and the economy.

Hydrocarbons can be toxic, mutagenic, carcinogenic, bioaccumulative and persistent in the environment. They can alter the physico-chemical and biological properties of water, reducing dissolved oxygen and available nutrients. They can also damage the structure and functioning of marine ecosystems, affecting biodiversity and productivity [2].

The entire food chain can be contaminated, transferring from lower to higher organisms, including humans. They can cause material and economic damage, affecting fishing activities, tourism and shipping.

2.2 Experimental Data from the Romanian Water Costal Area

The quality of costal water characterization programs will aim at estimating one or more static parameters defining the concentration of one or more parameters to be determined, its variations, or both, during a given period [3].

For example, the average or median indicates the overall trend of results, and the standard deviation indicates dispersion. The results may be required either for a research program for which no continuous control is required or in the framework of long-term control [3].

To compare the obtained values and determine the alert threshold, the reference values were used according to Order no. 756/1997 for the approval of the Regulation on environmental pollution assessment. The regulation regarding the classification of surface waters, annex B for: heavy metals (cadmium, copper), pesticides (sum of DDD/DDT/DDE), total polynuclear aromatic compounds (sum of PAH). Total polychlorinated biphenyl compounds (PCB amounts), mononucleic aromatic compounds (benzene, toluene, phenols) [3, 4].

2.3 Methodology and Results

2.3.1 Methodology

The TPH values for the period 2021–2022 were obtained in collaboration with the Romanian Waters Administration Dobrogea Litoral.

For the 2023 period, we took specific water samples from nearby areas, which were sent for testing to the accredited laboratory for PAH determinations in water [5].

For the period 2021–2023, we analyzed and compared the data received and those obtained with the values imposed by law, thus obtaining the chart for their standard deviation.

The calculation of the standard deviation was made to determine before performing the risk analysis, the degree of pollution in the chosen areas on the Romanian Black Sea Coast.

The Black Sea is an isolated ecosystem and very vulnerable to negative anthropogenic interventions.

2.3.2 *Results*

Tables 2.1, 2.2 and 2.3 contains input data for environmental risk analysis. They were used to calculate the standard deviation.

Table 2.1 Values for THP-2021

Section name-2021	Total hydrocarbons (dissolved or in emulsion) (μg/l)			Standard deviation
	MIN	MA	MAX	
Navodari—5 m	156	164	173	8,50
Navodari—20 m	144	195	268	62,32
Constanta North—shore	53,5	136,13	174	61,62
Constanta North—5 m	53,5	141,5	200	73,74

Table 2.2 Values for THP-2022

Section name-2022	Total hydrocarbons (dissolved or in emulsion) (μg/l)			Standard deviation
	MIN	MA	MAX	
Navodari—5 m	108	127	146	20
Navodari 20 m	120	140	160	36,17
Constanta North—shore	53,5	70,875	123	30
Constanta North—5 m	108	138	168	19

Table 2.3 Values for PAH -2023

Place of sampling		Navodari—5 m 08.08.2023	Navodari—20 m 08.08.2023	Constanta North—shore 08.08.2023	Constanta North—5 m 08.08.2023	Standard deviation value
Sample date						
Parameter	UM	Result	Result	Result	Result	
Naphthalene	μg/L	8,75	9,62	9,18	10,06	0,56
Acenaphylenes	μg/L	1,25	1,37	1,31	1,43	0,08
Acenaphthan	μg/L	1,25	1,37	1,31	1,43	0,08
Fluoren	μg/L	1,25	1,37	1,31	1,43	0,08
Phenanthrene	μg/L	1,25	1,37	1,31	1,43	0,08
Antracene	μg/L	1,25	1,37	1,31	1,43	0,08
Fluorantine	μg/L	1,25	1,37	1,31	1,43	0,08
Pyrenees	μg/L	1,25	1,37	1,31	1,43	0,08
Benzo (a) antracen	μg/L	1,25	1,37	1,31	1,43	0,08
Crisen	μg/L	1,25	1,37	1,31	1,43	0,08
Benzo (b) fluorantene	μg/L	1,25	1,37	1,31	1,43	0,08
Benzo (k) fluorantene	μg/L	1,25	1,37	1,31	1,43	0,08
Benzo (a) pyrene	μg/L	1,25	1,37	1,31	1,43	0,08
Indeno (1.2.3.cd) pyrene	μg/L	0,375	0,41	0,39	0,43	0,02
Benzo (g.h.i) perilene	μg/L	0,375	0,41	0,39	0,43	0,02
Dibenzo (a,h) antracene	μg/L	0,75	0,82	0,78	0,86	0,04
Amount 16 PAH	μg/L	25,25	27,7	26,51	29,03	1,63

2.3.2.1 The Variations of THP in the Period 2021–2022

The results of THP (Total Hydrocarbons) were presented in the period 2021–2022 on the coast of Constanța at a depth of 5 m from the coast in the following tables (Tables 2.1 and 2.2).

The PAH (Polycyclic Aromatic Hydrocarbons) results were presented in the period 2023 on the Constanța-North coast at 5 m depth from the coast in the following tables (Table 2.3).

2.3.2.2 The Standard Deviation

Figures 2.1, 2.2 and 2.3 shows for each period the comparison of reference values and standard deviation values for each area under review.

Following the analysis of the registration data and their comparison with the reference values according to the legislation in force, it was found that the analyzed area of the Romanian Black Sea Coast has a minor water pollution.

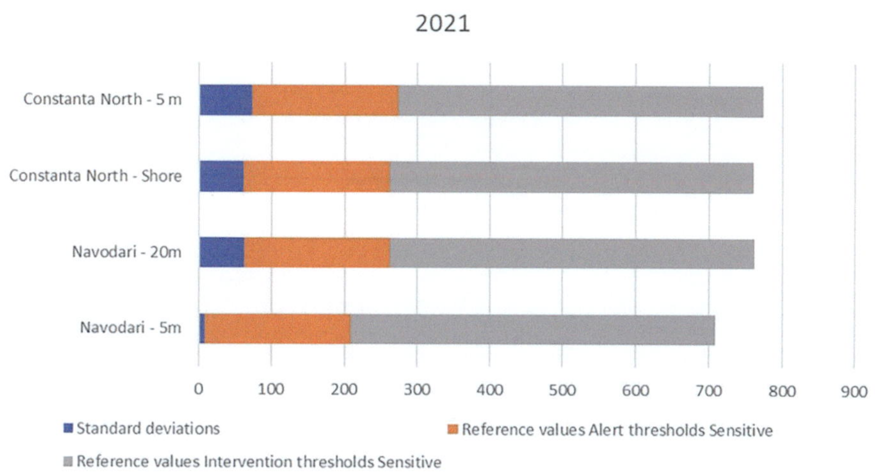

Fig. 2.1 Comparison of reference values and standard deviation values-2021

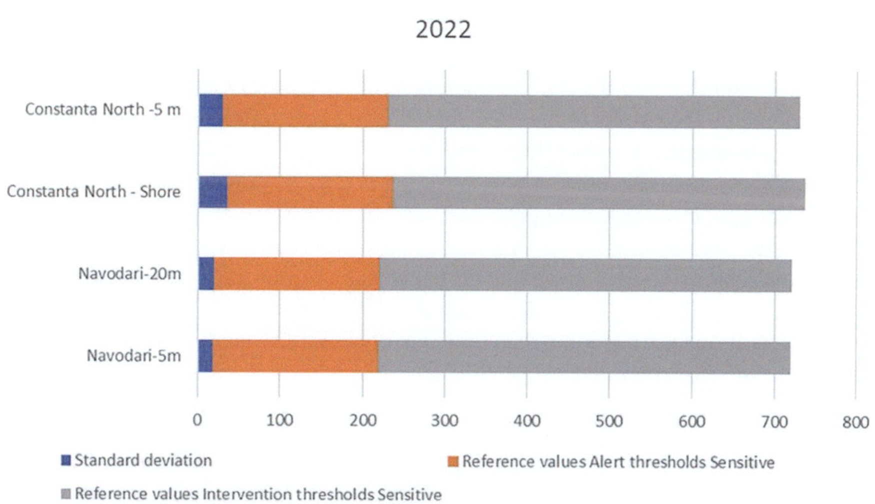

Fig. 2.2 Comparison of reference values and standard deviation values-2022

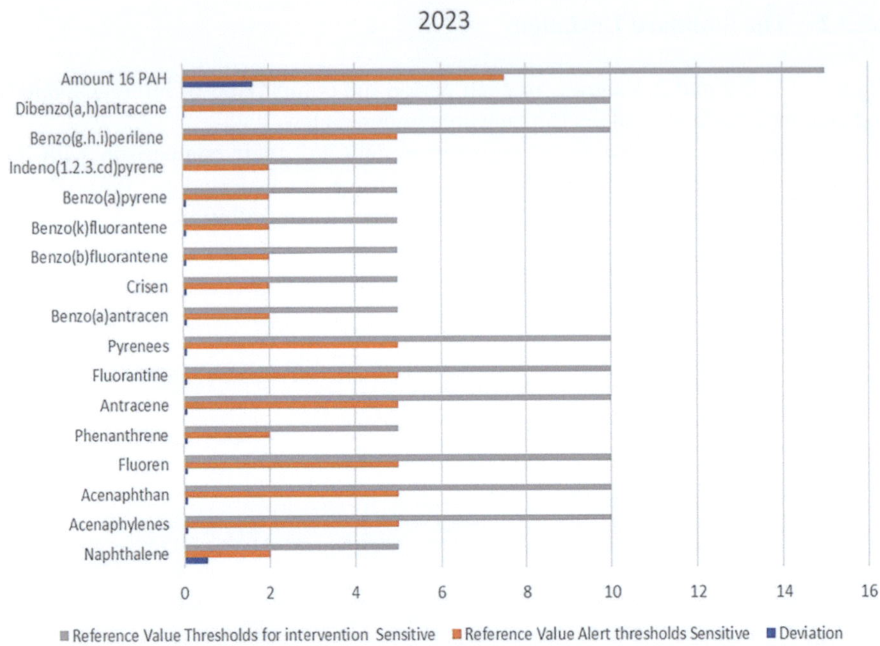

Fig. 2.3 Comparison of the standard deviation and the reference values for 2023

2.4 Conclusions

The results of the analyses indicate a stability of the environment in the Black Sea Coast area, with a low standard deviation in terms of hydrocarbon and sediment concentrations in water and soil. This suggests that the area is generally less affected by pollution and that soil quality parameters remain relatively constant [6].

In conclusion, the results of the experimental research indicate a favorable general state of the environment in the Black Sea Coast area, with a small standard deviation and a relatively polluted area. These findings are an important starting point for conservation and environmental protection efforts in this region [7, 8].

The results obtained in this paper indicate a favorable general state in the analyzed area, in conclusion, the anthropogenic influence has a minor impact on the Romanian Seaside.

These results are used as input data for environmental risk analysis on hydrocarbon pollution in the Romanian Black Sea Coast.

References

1. Encyclopedia of Romania.The Black Sea. https://enciclopediaromaniei.ro/wiki/Marea_Neagr%C4%83. Accessed 19 May 2025
2. Black Sea 21. https://mareaneagra21.weebly.com
3. ORDER NO. 756/1997. (n.d). Regulation on environmental pollution assessment
4. EN ISO/IEC 17025:2018. (n.d). General requirements for the competence of testing and calibration laboratories
5. ISO 18287. (n.d). Determination of polycyclic aromatic hydrocarbons (PAH). Gas chromatographic method with mass spectrometric detection (GC-MS)
6. Delgado, L., Kumzerova, E., Martynov, M.: Simulation of oil spill behavior and response operations in PISCES. WIT Trans. Ecol. Environ. **88**, 279–292 (2006)
7. Panaitescu, F.V., Panaitescu, M., Voicu, I.: Environmental Risk Concepts. Nautica Publishing House (2019)
8. Popescu, D.-M., Nistoran, D.E.: Surface Water Pollution with Petroleum Liquids

Chapter 3
Study of the Drinking Water Quality from Dambovita County

Ionica Ionita, Ana-Maria Hossu, Daniela Avram, Aurora Anca Poinescu, and Roxana Elena Margarit

Abstract The analytical evaluation of the main physicochemical parameters of drinking water from the selected areas in Dambovita county during the year 2023 it appears that all the parameters correspond to both national and international drinking water regulations. For drinking water samples from the Dambovita county, quality indicators were analyzed: pH, turbidity, salinity, oxidability, Cl^-, nitrites, total hardness, Ca^{2+} and Mg^{2+}, at the time of sampling, in real time using specific analysis methods.

Keywords Water · pH · Conductivity · Turbidity · Oxidability

3.1 Introduction

The natural composition of soil and rocks can introduce minerals, salts, and other contaminants into groundwater, affecting the quality of fountains waters from rural areas. Agricultural runoff and improper waste disposal can all pollute nearby water sources, compromising water quality.

I. Ionita · A.-M. Hossu (✉)
Department of Sciences and Advanced Technologies, Valahia University of Targoviste, Targoviste, Romania
e-mail: ana.hossu@valahia.ro

I. Ionita
e-mail: ionica.ionita@valahia.ro

D. Avram · R. E. Margarit
Department of Food Engineering, Valahia University of Targoviste, Targoviste, Romania

A. A. Poinescu
Department of Material Engineering, Valahia University of Targoviste, Targoviste, Romania

R. Tamas and C. A. Heiman (eds.), *Selected Papers from AToMS 2024*,
Lecture Notes on Multidisciplinary Industrial Engineering,
https://doi.org/10.1007/978-3-032-02452-7_3

Natural waters have an intake of mineral and organic substances, which are dissolved or in the form of naturally occurring colloids or suspended solid particles, which do not allow them to be used directly for human consumption or to produce food and pharmaceutical products.

Natural water (surface water and groundwater) is never pure, it undergoes continuous physical and chemical transformations, which greatly affect its composition. In its natural circuit, water dissolves a number of gases from the air (O_2, N_2, CO_2, H_2S, etc.) and then dissolves some of the soluble minerals on the earth's surface and through infiltration, enriching itself in mineral salts (chlorides, bicarbonates, sulphates, nitrates, phosphates, silicates, especially of the alkali metals Na^+, K^+ and alkaline earth metals Ca^{2+}, Mg^{2+}). In addition to inorganic materials, natural water also contains organic substances in dissolved or colloidal form.

The water potability criteria from ground or surface sources include several organoleptic, physicochemical and microbiological parameters, such as odour, colour, temperature, pH, conductivity, turbidity, Cl^-, total hardness (TH), nitrites, coliform bacteria, etc. [1].

This study presents the results of drinking water analysis in Dambovita county through the prism of physicochemical and microbiological indicators, according to the standards in force, especially for its use in human consumption and in the food industry (raw and auxiliary material for obtaining food products, obtaining technological steam, ice, etc.) according to Law 458/2002 and STAS 1342–84.

The groundwater sources in Dambovita county, Romania (Fig. 3.1) depend on the degree of permeability as well as on the thickness and extension of the rocks that store and release them to the surface.

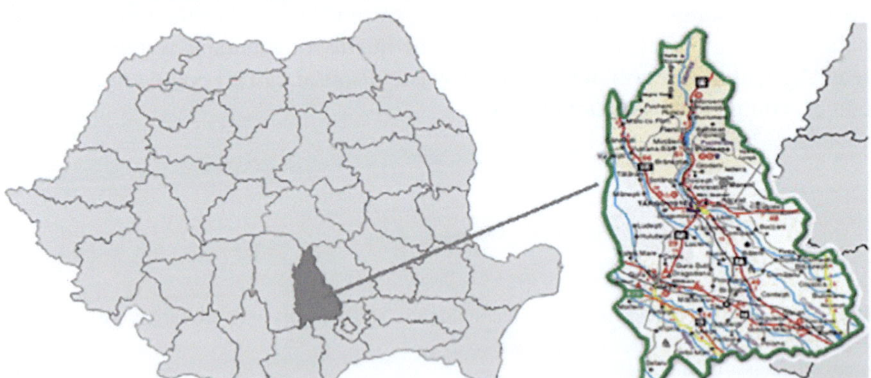

Fig. 3.1 Location map of the study county

3.1.1 Analysis of Drinking Water from Rural Areas

Physical, chemical and epidemiological studies on the effects of environmental pollution of drinking water demonstrate a strong influence of the harmful effects of chemical and microbiological pollutants at young ages for people. The quality indicators pH, NO_2^- turbidity, salinity, oxidability, chlorides, total hardness (TH), $Ca^{2+}$and Mg^{2+} were analysed for drinking water samples from fountains at the time of sampling in real time using specific analytical methods according to current international standards (Table 3.1 and 3.2).

Figure 3.2 shows the water sampling points: Baleni (44°49′30.6"N 25°38′57.2"E), Cojasca (44°43′30.9"N 25°51′15.9"E) and Brezoaele (44°34′47.7"N 25°45′02.2"E).

3.2 Results and Discussion

A total of 50 samples were collected during one day in all seasons of the year 2023, using clean polyethylene bottles for the determination of physicochemical parameters (limits allowed for Romania are according to Ordinance no. 7/18.01.2023).

Electrical conductivity is an important parameter in evaluating the quality of drinking water. It measures the water's ability to carry an electric current, which is influenced by the presence of dissolved salts and minerals. The portable conductivity meters are used to measure the electrical conductivity of drinking water samples and detect the concentration of dissolved ions. The WTW Portable Conductivity Meter ProfiLine Cond 3110 has been used to determine conductivity, salinity and temperature (Table 3.1). The acceptable range of electrical conductivity in drinking water varies depending on local regulations and water quality standards. Generally, drinking water with an electrical conductivity between 50 to 1500 μS/cm is considered safe for consumption.

The water samples (Fig. 3.3) show that the conductivity of the water in Dambovita county varies between 538 and 720 μS/cm. These results show that all the analyzed waters have an electrical conductivity in accordance with the Romanian standards (maximum 2500 μS/cm).

The electrical conductivity values are between 500 and 1000 μS/cm and are therefore, according to this classification, potable waters [2].

The pH determination was carried out with an InoLab WTW 720 pH meter (according to SR EN ISO 10523:2012), its values are in the range of 6.7 and 7.4 in most water samples. The turbidity (based on the nephelometric technique) which indicates the cloudiness or suspended particles in the water has been accomplished with a Turb-550 turbidimeter in accordance with SR EN ISO 7027:2001.

The determination of chloride content (Mohr's method) in water samples (SR ISO 9297:2001) from Dambovita county values well below the minimum detectable concentration (250 mg/L) mentioned in the Romanian legislation. Mohr's method (precipitation titration) uses 0.02N $AgNO_3$ as the titrant and a quantity of 5% K_2CrO_4

Table 3.1 The quality indicators obtained during the year 2023 of drinking water

Quality indicator/ reference standard	MCL drinking water	Annual season	Baleni	Cojasca	Brezoaele
Temperature (°C)/ SR EN 1622:2000	–	Spring	15.89	15.11	15.59
		Summer	17.89	16.97	18.01
		Autumn	16.04	16.34	16.22
		Winter	14.01	14.72	14.87
pH/ SR EN ISO 10523:2012	6,5–9,5	Spring	7.4	7.1	7.2
		Summer	7.5	7.0	6.9
		Autumn	7.3	6.9	6.7
		Winter	7.1	7.0	7.0
Conductivity [μS/cm]/ SR ISO 27888:1997	2500	Spring	538	581	645
		Summer	670	604	720
		Autumn	650	615	713
		Winter	587	598	690
Salinity [‰]/ SR ISO 27888:1997	–	Spring	0.2	0.2	0.3
		Summer	0.3	0.4	0.5
		Autumn	0.2	0.3	0.4
		Winter	0.2	0.2	0.3
Turbidity [NTU]/ SR EN ISO 7027–1:2016	4	Spring	0.34	0.76	0.85
		Summer	0.19	0.64	1.56
		Autumn	0.06	1.04	0.69
		Winter	0.04	0.83	0.21
Oxidability [mg O_2/L]/ SR EN ISO 8467:2001	5	Spring	0.54	0.58	0.63
		Summer	0.61	0.64	0.79
		Autumn	0.59	0.59	0.80
		Winter	0.82	0.61	0.74
Cl^- [mg/L]/ SR ISO 9297:2001	250	Spring	17.46	15.36	18.34
		Summer	19.55	16.32	19.98
		Autumn	18.07	15.96	18.45
		Winter	16.98	14.68	18.02
NO_2^- [mg/L]/ SR EN 26,777:2002/ C91:2006	0.5	Spring	0.011	0.013	0.023
		Summer	0.013	0.015	0.033
		Autumn	0.012	0.012	0.025
		Winter	0.010	0.011	0.019

Table 3.2 Magnesium and calcium ions concentration in waters from Dambovita county

Ion concentration	Annual season	Baleni	Cojasca	Brezoaele
Ca^{2+} [mg/L]	Spring	78	66	71
	Summer	83	70	73
	Autumn	85	69	72
	Winter	79	65	69
Mg^{2+} [mg/L]	Spring	31	28	28
	Summer	35	31	30
	Autumn	35	30	29
	Winter	30	26	27

Fig. 3.2 The study area of drinking water quality indicators in Dambovita county

as the indicator to determine the concentration of chloride in water sample from fountains. The chloride was precipitated as AgCl, and the titration was finalized when the color of the indicator (K_2CrO_4) changed (from yellow to yellow-pink).

Oxidability (determined through the permanganate index parameter, analyzed according to the $KMnO_4$ oxidation method) is a crucial factor in the assessment of drinking water quality [3]. It measures the presence and concentration of oxidizing

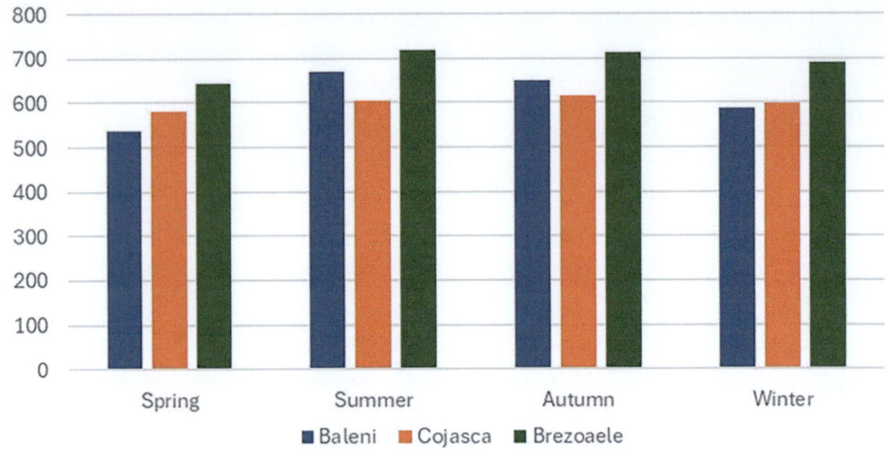

Fig. 3.3 Conductivity variation of samples water [μS/cm]

substances, which can significantly impact the water's taste and odour. Dissolved oxygen levels, mineral content, and other chemical factors can impact on the rate of oxidation, affecting the taste, clarity, and quality of the water. The oxidability value of samples from natural water sources is kept below the value of 1 mg O_2/L corresponding to the maximum contamination level required by legislation.

The values of the oxidability index are below the maximum allowed limit, 5 mg O_2/L water, according to the Romanian standards, and the highest oxidability was found in the water from Brezoaele village 0.80 mg O_2/L (autumn), the oxidability values being conditioned by the amount of natural biodegradable organic substances.

Nitrite is an important water quality parameter that must be monitored in drinking water from fountains. Elevated nitrite levels can pose health risks, making this a crucial consideration for fountain owners and water authorities. Fertilizer runoff and manure applications can increase nitrate concentration, which convert to nitrite. Maintaining nitrite within regulatory limits is essential for providing safe, potable drinking water, also natural mineral deposits can influence nitrite levels in groundwater. The determination of nitrite content is important because it has harmful effects on the human body. It was determined spectrophotometrically by the Griess method. Direct spectrophotometric determination is a widely used analytical technique for the quantitative measurement of various chemical species in a sample.

Proper sample preparation is determinant to ensure accurate and reliable results in the direct spectrophotometric measurement of nitrite concentration (520 nm) using UV–vis spectrophotometer (Shimadzu UV-1900i) according to SR EN 26,777 ISO 6777–2002 [4]. The determination of nitrite can be affected by the presence of other compounds in the sample, such as nitrate, chloride, or organic matter, which may interfere with the colour development or absorbance measurement. Proper sample preparation, including filtration, dilution, and pH adjustment, is crucial to ensure accurate results and minimize matrix effects. This method offers a simple, rapid, and cost-effective approach for the accurate determination of nitrite concentrations in various samples. Analysis of nitrite test results shows that this parameter is below the permissible limit for all water samples in the county (<0.5 mg/L).

According to Law no. 458 of 8 July 2002 on the quality of drinking water, the permissible amount of Ca^{2+} is up to 100 mg/L or exceptionally up to 180 mg/L. In the case of Mg^{2+} ions in drinking water it is 50 mg/L and in exceptional cases—up to 80 mg/L, total hardness maximum 20^0D. In most of the drinking water samples. Calcium concentration values are within the accepted limit of potability (Table 3.2).

Figure 3.4 shows the concentration values of calcium and magnesium ions in drinking water analysed from water samples in Dambovita county. A higher value of calcium ion concentration is noticed in the water from Baleni (85 mg/L in autumn period). In all water samples analysed, the concentration of magnesium ions is below the maximum limit of 80 mg/L.

Evaluating water hardness is crucial to avoid mineral buildup and crusts deposit inside the pipes, appliances, and irrigation systems. The total hardness of fountain water samples are determined by the complexometric method. This involves collecting representative water samples, preserving them under controlled conditions and then performing a quantitative laboratory analysis using the Na_2EDTA complexometric titration method. For all analysed samples, the total hardness varies between 15–20^0D (German degrees), which represents a water with increased hardness (Table 3.3), because in our country, 85% of drinking water sources are impaired, having a high level of salts and calcium and magnesium minerals.

The microbiological testing for water quality

Monitoring parameters like coliform bacteria helps identify potential microbial contamination that could pose health risks. From the microbiological point of view,

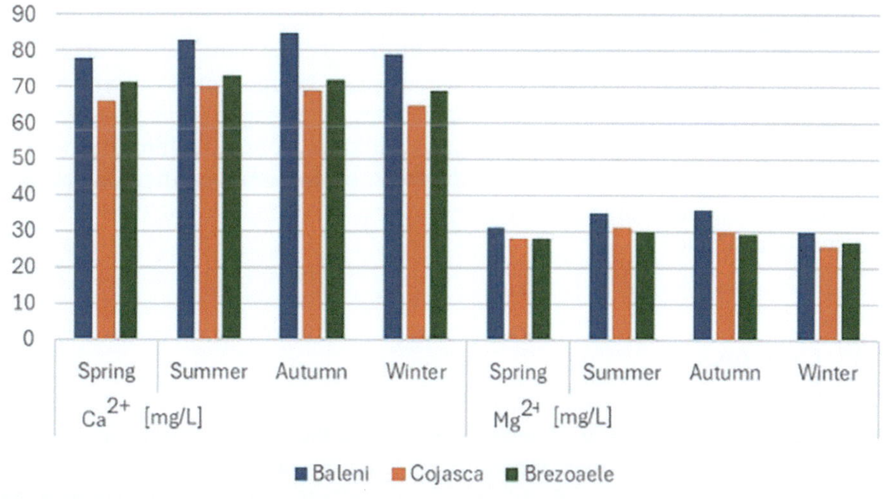

Fig. 3.4 Magnesium and calcium ions concentration in drinking water from Dambovita county

Table 3.3 The total hardness of water samples

TH [^{0}D] Method of analysis SR ISO 6059	Annual season	Baleni	Cojasca	Brezoaele
	Spring	18	15.7	16.4
	Summer	19.7	16.9	17.1
	Autumn	19.9	16.5	16.7
	Winter	17.9	15.3	15.9

Table 3.4 The value of coliform bacteria in water samples

Area of the county	MCL drinking water	Coliform bacteria [CFUs/100 mL/250 mL drinking water]
Baleni	0 CFUs	0
Cojasca		0
Brezoaele		0

coliform bacteria are not allowed. That is why water samples were tested to determine whether drinking water is safe. If they appear in a water sample, it often indicates contamination with fecal source pathogens. Most coliform bacteria are conditionally pathogenic, only some strains of Escherichia coli can cause quite serious illness.

The presence of coliform bacteria in water generally indicates recent pollution. For coliform bacteria, Ordinance 7/18.01.2023 on drinking water quality stipulates for their absence/100 mL/250 mL water. From the analysis of the test results for coliform bacteria, it appears that this parameter is within the legal limits for all samples considered for fountains water in Dambovita county (Table 3.4). More precisely, coliform bacteria were not identified in the water samples analysed.

The determination method is based on filtration through sterilizing membrane followed by cultivation on chromogenic coliform agar medium and calculation of the number of target microorganisms in the sample (Chromogenic Coliform Agar).

The SR EN ISO 9308–1:2015 standard includes the principle of the measurement method, requirements for the equipment and glassware used, culture media and reagents, sampling method, working method (sample preparation, filtration, incubation and differentiation), expression of results and method of writing the test report and quality assurance.

3.3 Conclusions

Advantages of physicochemical analysis provides a thorough examination of water quality, assessing a wide range of parameters to ensure drinking water meets safety standards. This analysis can identify potential contaminants or issues before they become a public health concern and is essential for demonstrating compliance with stringent drinking water regulations and guidelines.

Regular testing of key water quality parameters demonstrates ongoing compliance with mandated limits for contaminants, disinfectants, and other critical factors.

At the end of this study, which focused mainly on the assessment of the quality of drinking water in Dambovita county from the three points, it was found that all the parameters analysed meet both national and international drinking water regulations.

The main quality indicators considered critical for water were investigated: temperature, pH value, turbidity, oxidability, TH, Cl^-, NO_2^-. None of the quality indicators showed exceedances of the admissible concentration limit and the drinking water in the groundwater in Dambovita county meets the drinking water quality standards without any risk to the health of consumers.

The water consumption by humans must be safe, so fountains should be at least 50 m away from possible sources of pollution: latrines, animal pens, rubbish and fertilizer dumps, because they stay in the soil for a long time and can pollute the water. Fountains must be monitored by water analysis (identification and evaluation of different physicochemical parameters) twice a year to have drinking water that is safe for consumption, compliant both qualitatively and quantitatively. Physicochemical and microbiological analysis of drinking water is a required tool for ensuring public health and maintaining water quality, although it has some inherent limitations.

Current primary analysis involves only the determination of basic physicochemical and microbiological indicators, cannot detect all potential contaminants, and some harmful substances may be missed. In addition, determinations can be time-consuming and costly, posing a challenge for regular and comprehensive water quality monitoring.

References

1. Liu, Y., Zheng, B.H., Fu, Q., Wang, L.J., Wang, M.: The selection of monitoring indicators for river water quality assessment. Procedia Environ. Sci. (2012). https://doi.org/10.1016/j.proenv.2012.01.013
2. Rodier, J., Legube, B., Merlet, N., Brunet, R.: Water analysis-10th edition. Natural Waters, Waste Waters, Sea Water, Technical and Engineering: Dunod France (2009)
3. Sullivan, P.J., Agardy, F.J., Clark, J.J.: The Environmental Science of Drinking Water. Elsevier, Amsterdam (2005)
4. Bajic, S.J., Jaselskis, B.: Spectrophotometric Determination of Nitrate and Nitrite in Natural Water and Seawater (1985). https://doi.org/10.1016/0039-9140(85)80038-2

Chapter 4
Risk Assessment Using New Environmental Simulation Tools

Florentina A. Vasilică, Carmen A. Safta, Mariana Panaitescu, Fănel V. Panaitescu, and Cristina Sofroniei

Abstract In Constanta and Navodari region, close to the Black Sea, Romanian coastal line, there are several hydrological risks associated with the specific hydrological characteristics of the area and human activities. Assessment and management of human impact are crucial aspects in efforts to promote and maintain Good Ecological Status in the Black Sea. It is essential to adopt appropriate measures to reduce pollution, conserve marine resources and protect natural habitats to ensure the health and sustainability of the marine ecosystem in this region. This paper presents risk assessment using the New Environmental Assessment Tools software. The results show the good ecological condition of the analyzed ecosystems and the anthropogenic impact on the environmental factors in the study locations. These results are useful for establishing the risk matrix and the global pollution index.

Keywords Ecosystem · Black Sea · Assessment · Tool · Status · Risk · Sensitivity · Statement · Severity · Vulnerability

4.1 Introduction

The Black Sea, one of the most important seas in the world, is a vital ecosystem for its surrounding region, providing essential natural resources and supporting regional biodiversity. However, anthropogenic pressures on water and soil quality in North Constanta and Navodari have increased significantly in recent decades, endangering the health of this fragile environment [1].

The assessment and monitoring of water quality thus become crucial priorities for the preservation and protection of the Black Sea ecosystem. In this context, the use

F. A. Vasilică (✉) · C. A. Safta
National University of Science and Technology POLITEHNICA Bucharest, Bucharest, Romania
e-mail: vasilica.florentina16@gmail.com

M. Panaitescu · F. V. Panaitescu · C. Sofroniei
Faculty of Naval Electromechanics Engineering and Environmental Protection in Industry, Constanta Maritime University, Constanta, Romania

R. Tamas and C. A. Heiman (eds.), *Selected Papers from AToMS 2024*,
Lecture Notes on Multidisciplinary Industrial Engineering,
https://doi.org/10.1007/978-3-032-02452-7_4

of modern technologies, such as NEAT + (Nexus Environmental Assessment Tool) software, becomes imperative for the effective understanding and management of risk factors [2].

The purpose of this work is to investigate and evaluate the water quality in the Constanța Nord and Năvodari area, with an focus on the identification and analysis of risk factors that directly or indirectly affect this environment.

4.2 Risk Assessment Using New Environmental Simulation Tools

4.2.1 Methodology

The calculation of the Good Ecological Condition (GSE) for the Black Sea involves a wide range of indicators, depending on the specific evaluation objectives and available resources, such as [3]:

- Water quality: Water quality is a fundamental aspect of the assessment of the Good Ecological Condition (GSE) in the Black Sea and in any other aquatic ecosystem. The concentration of Chemicals measures the presence and concentration of various chemicals in the water. Chemicals can include heavy metals, pesticides, organic substances, toxic substances or chemical pollutants. High concentrations can have a negative impact on aquatic organisms and human health. The following tables will show the values of the concentrations of chemicals between 2019–2023. Tables 4.1, 4.2 and 4.3 presents the values of polynuclear aromatic hydrocarbons (PAH) for the years 2019, 2020, 2023, and Table 4.4 presents the values of total hydrocarbons (TPH) [4–8].
- Diversity of Animal and Plant Species: Species diversity is a key indicator of the ecological health of an ecosystem. It is important to note that the Black Sea has significant biological productivity, despite these limitations [9]. A considerable number of marine species have been identified in the Black Sea, including about 3,774 species, most of which are invertebrates and plants. Although macroscopic biodiversity is relatively well documented, there is also a rich diversity of microscopic organisms, such as viruses, bacteria, fungi, algae and protozoa, as well as, that still requires detailed scientific research to be fully understood [10].
- Air Quality: Air quality can be influenced by various factors, including industrial activities, road traffic, agricultural exploitation and weather conditions. Monitoring air quality in this region can provide essential information about air pollution levels and the impact on human health and the environment [11].
- Soil Condition: Soil condition is another important aspect of the assessment of the Good Ecological State (GSE) in the Black Sea. Elements such as oxygen and nutrient levels, chemical and biological contaminants, sediments and anthropogenic pressure can influence the condition of the soil [10].

Table 4.1 Value PAH-2019

No crt	PAH component (ng/l)	2019			
		CT 01	CT 02	MID 02	CT 03
1	Naphtalin	1,75	1,68	3,05	2,10
2	Acenaphthylene	19,85	19,87	20,39	20,08
3	Acenaphthens	10,98	10,26	12,74	10,76
4	Fluorene	16,99	16,95	17,16	17,02
5	Phenanthrene	19,05	18,77	19,15	18,74
6	Anthracene	26,20	26,20	20,20	27,52
7	Fluoranthene	25,22	25,27	25,29	25,22
8	Pyreneur	26,71	26,82	26,65	26,63
9	Benzo(a)anthracene	32,33	32,41	32,31	32,32
10	Crisens	27,04	27,24	27,03	27,03
11	Benzo(b)fluoranthene	34,19	34,30	34,20	34,26
12	Benzo(k)fluoranthene	29,81	29,89	29,88	29,93
13	Benzo(a)pyren	36,01	36,04	35,99	35,94
14	Indeno	N.D	36,15	N.D	N.D
15	Dibenzo(a,h)anthracene	N.D	N.D	N.D	N.D
16	Benzo(ghi)perylene	N.D	27,88	N.D	27,69
	Total amount (ng/l)	306,13	369,73	304,04	335,24

- Human impact: Human impact is an important factor in the assessment of the Good Ecological State (SEB) of the Black Sea. The evaluation and management of human impact are crucial aspects in the efforts to promote and maintain a Good Ecological State in the Black Sea. It is essential to adopt appropriate measures to reduce pollution, conserve marine resources and protect natural habitats to ensure the health and sustainability of the marine ecosystem in this region [12, 13].
- Risks: Geological: In the Constanta and Navodari region near the Black Sea, there are certain geological risks associated with specific geological features and human activities in the area. Hydrological: In the Constanta and Navodari region, close to the Black Sea, there are several hydrological risks associated with the specific hydrological characteristics of the area and with human activities [14].
- Climate change: Climate change is one of the biggest challenges of the XXI century and has significant impacts on the entire planet, including the Constanta and Navodari region. These climate changes are mainly caused by human activities, in particular greenhouse gas emissions from the burning of fossil fuels, deforestation and other industrial activities. To meet these challenges, it is essential that local communities, government authorities and other stakeholders take action to adapt and mitigate climate change. These measures may include reducing greenhouse gas emissions, protecting and restoring natural ecosystems, sustainable urban planning and creating climate-resilient infrastructures [15].

Table 4.2 Value PAH-2020

No crt	PAH component (ng/l)	2020			
		CT-02	SG-01	CT-03	MID-02
1	Naphtalin	18.82	17,56	20,13	16,82
2	Acenaphthylene	15,48	15,31	14,49	14,46
3	Acenaphthens	12,97	17,19	18,45	17,17
4	Fluorene	20,94	21,07	20,44	20,67
5	Phenanthrene	38,13	39,00	36,68	37,47
6	Anthracene	22,22	22,16	22,07	22,22
7	Fluoranthene	21,13	21,22	20,92	20,73
8	Pyreneur	21,39	21,69	21,07	20,74
9	Benzo(a)anthracene	33,24	33,29	N.D	N.D
10	Crisens	19,28	19,30	N.D	N.D
11	Benzo(b)fluoranthene	22,25	22,32	22,15	22,11
12	Benzo(k)fluoranthene	22,31	22,38	22,21	22,17
13	Benzo(a)pyren	26,90	26,97	26,83	26,83
14	Indeno	N.D	N.D	N.D	N.D
15	Dibenzo(a,h)anthracene	36,44	36,32	36,38	36,29
16	Benzo(ghi)perylene	28,86	29,12	N.D	N.D
	Total amount (ng/l)	341,54	364,90	281,82	277,68

- Energy and Natural Resources: Constanta and Navodari region is significant in terms of energy and natural resources, due to the diversity of landscapes and the natural wealth of the region [16].

4.2.2 Results

The use of NEAT + (Nexus Environmental Assessment Tool) software in risk assessment brings multiple advantages and contributes to a deeper understanding of the impact of human activities on the environment in the area of Constanta North and Navodari from the proximity of the Black Sea [2].

Through NEAT+, we can carry out a detailed assessment of the various risk factors that may affect the quality of water and soil in this region. The software enables the analysis and integration of data from diverse sources, including geospatial information, weather data, water and soil quality monitoring reports, as well as socio-economic data.

In Fig. 4.1, the input data for the analysis are presented.

The results of the NEAT + analysis, Fig. 4.2, show the impact of hydrocarbon pollution in the Romanian coastal area presented in Table 4.5.

Table 4.3 Value PAH-2023

No crt	PAH component (ng/l)	2023			
		CT-02	SG-01	CT-03	MID-02
1	Naphtalin	8,75	9,63	9,19	10,06
2	Acenaphthylene	1,25	1,38	1,31	1,44
3	Acenaphthens	1,25	1,38	1,31	1,44
4	Fluorene	1,25	1,38	1,31	1,44
5	Phenanthrene	1,25	1,38	1,31	1,44
6	Anthracene	1,25	1,38	1,31	1,44
7	Fluoranthene	1,25	1,38	1,31	1,44
8	Pyreneur	1,25	1,38	1,31	1,44
9	Benzo(a)anthracene	1,25	1,38	1,31	1,44
10	Crisens	1,25	1,38	1,31	1,44
11	Benzo(b)fluoranthene	1,25	1,38	1,31	1,44
12	Benzo(k)fluoranthene	1,25	1,38	1,31	1,44
13	Benzo(a)pyren	1,25	1,38	1,31	1,44
14	Indeno	0,38	0,41	0,39	0,43
15	Dibenzo(a,h)anthracene	0,38	0,41	0,39	0,43
16	Benzo(ghi)perylene	0,75	0,83	0,79	0,86
	Total amount (ng/l)	25,25	27,78	26,51	29,04

Table 4.4 Value TPH in period 2021–2022

TPH	2021				2022			
(ng/l)	164,00	195,00	136,13	141,50	127,00	140,00	70,88	138,00

4.3 Conclusions

The evaluation and monitoring of the water quality in Constanta North and Navodari are crucial for the protection and conservation of the Black Sea ecosystem. Anthropogenic pressures on this fragile environment have increased significantly in recent decades, endangering its natural health and balance.

In this paper was use NEAT + software which demonstrate to be an effective and integrated solution for risk assessment and management of the impact of various human activities on the marine and terrestrial environment in this region. Through the detailed analysis of risk factors and the use of advanced modeling and simulation tools, NEAT + provides the opportunity to make informed decisions and implement appropriate preventive and corrective measures to protect the environment.

In conclusion, it is imperative to pay special attention to the quality of soil and water in the Navodari and North Constanta area. The environmental managing efforts

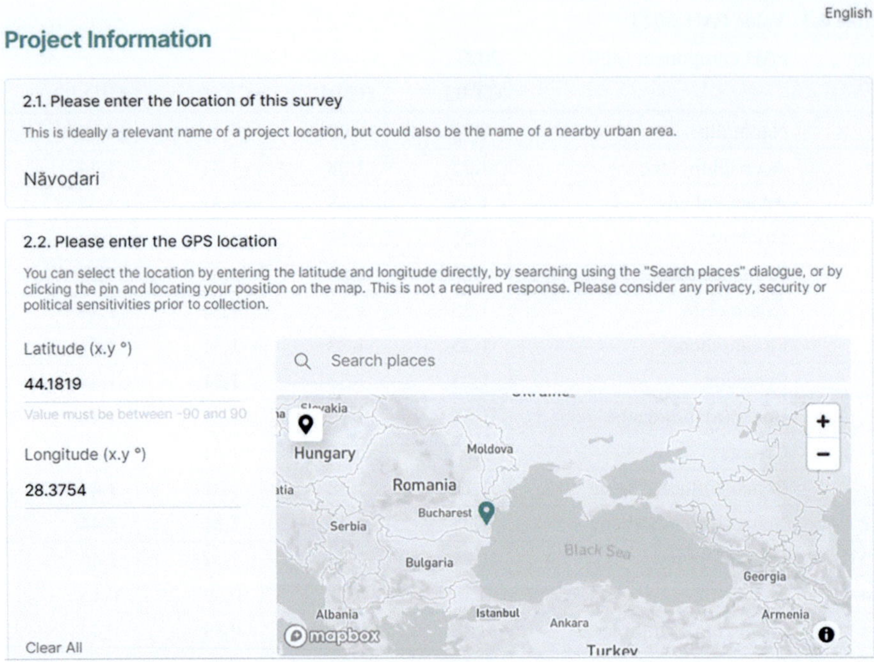

Fig. 4.1 Determination of the studied area

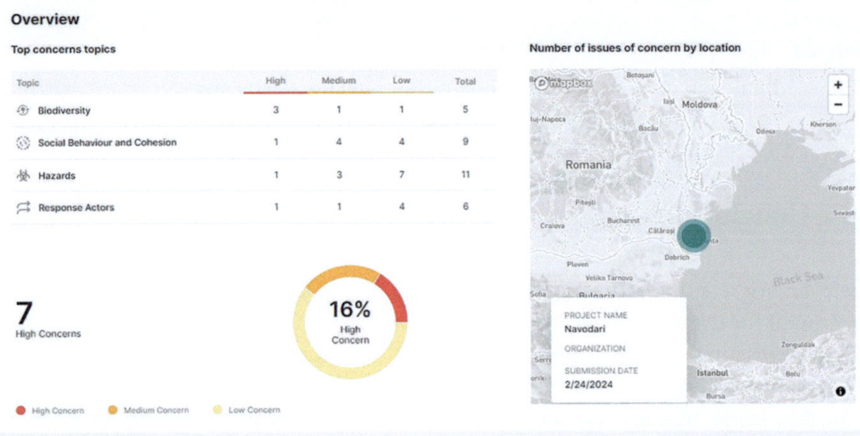

Fig. 4.2 NEAT + analysis results

Table 4.5 Impact of hydrocarbon pollution in Romanian coastal area

The degree of risk	Environmental impact
High impact risks	The environment which the population depend (local or remote) has high biodiversity value. Vulnerable and/or rare flora and fauna may be vulnerable
	There is low capacity to manage surface water drainage/wastewater. Environmental sanitation and disease transmission may be an issue
	The built environmental may not be resistant to industrial/technological hazards or pollution, which coul lead to loss of human life, insfrastructure and destruction of habitats
Moderate impact risks	Protection concerns may be exacerbated by environmental stresses (scarcity of resources, pollution, disease, degraded environment, deforestation, crop failures etc.)
	There is low capacity to manage sewerag and faecal sludge. Environmental sanitation and disease transmission may be an issue
	Traditional knowledge and practices may benefit or harm the environment
Low impact risks	Rates of deforestation may exceed regeneration capabilities
	Buit environment may not be resilient to geohazards, leading to loss of human life, infrastructure and destruction of habitats
	The area may be vulnerable to changing average and extreme temperatures due to climate change and urban areas acting as local heat sinks. It may also be vulnerable to climate related warming releasing GHGs from soil, affecting ecosystems and infrastructure [17, 18]

requires innovative solutions from all actors involved to find the best solution that could be able to be applied.

References

1. Black Sea 21. https://mareaneagra21.weebly.com
2. NEAT (Version 1.2).: Nested Environmental Status Assessment Tool Manual (n.d.)
3. International Organization for Standardization.: *SR ISO 31000: Risk management–Guidelines* (2018)
4. STAS 1242/4–85.: Geotechnical investigations through boreholes drilled in soils (1985)
5. International Organization for Standardization.: ISO 18400–104: *Soil Quality–Sampling* (2016)
6. International Organization for Standardization.: *ISO 18400–201: Soil Quality–Sampling—Physical Preparation* (2018)
7. International Organization for Standardization.: *ISO 5667–1: General Guide for Establishing Sampling Programs and Techniques* (2004)
8. International Organization for Standardization. (1995). ISO 5667–12: General guide for bottom sediment sampling.
9. Air Quality Index.: https://aqicn.org/city/romania/constanta/navodari/dc-86
10. Ministry of Environment.: Order no. 756 of November 3, 1997, for the approval of regulations on environmental pollution assessment (1997)
11. Micu, D.: *Romanian Marine Habitats of European Interest* (2007)

12. Năvodari Municipality.: https://www.primaria-navodari.ro/date-geografice WWF Report. http://d2ouvy59p0dg6k.cloudfront.net/downloads/partea_2_1.pdf
13. Red Book of the Black Sea.: https://ro.scribd.com/document/413029465/Cartea-Rosie-a-Marii-Negre
14. Romanian Encyclopedia.: https://enciclopediaromaniei.ro/wiki/Marea_Neagr%C4%83
15. European Statistical Portraits.: https://circabc.europa.eu/webdav/CircaBC/ESTAT/regportraits/Information/ro023_geo.htm
16. Delgado, L., Kumzerova, E., Martynov, M.: Simulation of oil spill behavior and response operations in PISCES. WIT Trans. Ecol. Environ. **88**, 279–292 (2006)
17. Popescu, D.-M., Nistoran, D.E.: *Surface Water Pollution with Petroleum Liquids*
18. Panaitescu, F.V., Panaitescu, M., Voicu, I.: *Environmental Risk Concepts*. Nautica Publishing House (2019)

Chapter 5
Consequences of a Major Crude Oil Spill in the Territorial Sea

Ionut Voicu, Mariana Panaitescu, and Fanel-Viorel Panaitescu

Abstract Regarding the political and military situation in the Black Sea Region, the authors propose to carry out a study on the environmental impact in the event of a major accidental spill of crude oil on the surface of the Black Sea water as a result of the explosion of a marine mine near an oil tanker. To carry out the study, we use the PISCES II Emergency Simulator, which is specially designed to be used in the process of training the personnel, from the management of the command center and operational center. Following the simulation, we will obtain data and information regarding the evolution of the crude oil spill over time, as well as the time period that the competent authorities have at their disposal to take all necessary measures to avoid pollution of the shoreline.

Keywords Crude oil spill · Pollution · Environmental impact · Black sea · Pisces II emergency simulator

5.1 Introduction

In the context of international situation generated by the Russian military aggression in Ukraine, a situation that led to a significant increase of naval traffic in Romanian ports on the Black Sea and beyond, the risk of events leading to major pollution in the territorial sea has increased exponentially, the tendency being for it to remain high in the following period as well.

I. Voicu (✉) · M. Panaitescu · F.-V. Panaitescu
Department of Mechanical and Environmental Engineering Sciences, Faculty of Naval Electromechanics, Constanta Maritime University, Constanta, Romania
e-mail: ionut.voicu@cmu-edu.eu

R. Tamas and C. A. Heiman (eds.), *Selected Papers from AToMS 2024*,
Lecture Notes on Multidisciplinary Industrial Engineering,
https://doi.org/10.1007/978-3-032-02452-7_5

5.2 Details About the Incident

Before entering to fairway of Constanta harbor, an oil tanker, on the move, is caught in an explosion of a sea mine, located in its vicinity.

As a result of sea mine explosion, which is not follow by a fire inside of the oil tanker, is made a perforation of the cargo hold, located in the starboard side, to the stern, next to the waterline, thus allowing that a quantity of crude oil to flow freely through the hole produced in the cargo hold of the ship.

Immediately, the incident is reported by the master of the oil tanker to the port authorities, who order that until the ships and depollution equipment of competent authorities arrive, the ship's crew should take all measures to limit the crude oil pollution of the sea water.

Following the explosion, the oil tanker goes into blackout, and its master orders the crew to start the anchoring procedure urgently. After fixing the problem that generated the blackout, the master of the oil tanker takes urgent measures to limit the pollution by transferring the crude oil from the damaged cargo hold to another cargo hold.

Approximately five hours after the explosion, following the measures taken by the crew, the flow of crude oil on the surface of the sea water is stopped. The amount of crude oil spilled during this period of time is approximately 920 tons.

5.3 Description of the Simulation Method

In order to determine the impact on the environment, due to the accidental flow of crude oil on the surface of the Black Sea water, following the incident described above, is used the Pisces II Emergency Situations Simulator, located in the Department of Mechanical and Environmental Engineering Sciences [3].

The Pisces II Simulator is specifically designed to be used in the process of training personnel within the competent authorities, and not only, who carry out practical activities of locating, limiting spreading and collecting oil products spilled on the surface of sea water [4].

The mathematical model used by Pisces II Emergency Situations Simulator takes into account the following aspects [2]:

- the main physic-chemical processes associated with the evolution of the oil on the water surface, such as: scattering, evaporation, dissolution, dispersion, emulsification, sedimentation and viscosity variation;
- the characteristics and properties of the oil, such as: density, water content, freezing point, flash point;
- environmental factors: shoreline, sea currents, sea state, sea water density, air/water temperature, ice, protected areas

5.4 Data Used to Perform the Simulation

The incident takes place at a distance of approximately six nautical miles across Cape Tuzla, in Romanian territorial sea. Hydrometeorological parameters used to perform the simulation are considered to be constant throughout the period for which the simulation is performed, as follows:

- sea water temperature: 288.15 K;
- air temperature: 285.15 K;
- sea water density: 1,015 kg/m^3;
- sea state: 1 m;
- speed/direction of the current: 0.5 m/s whit 280°;
- speed/direction of the wind: 1 m/s whit 270°.

We assume that crude oil is "Arabian Light" type, and has the following physical parameters, as follows:

- density: 858 kg/m^3;
- surface tension: 0.0168 N/m;
- kinematic viscosity: $1.63 \cdot 10^{-5}$ m^2/s;
- maximum water content: 87%;
- pour point: - 220.15 K;
- flashpoint: - 253.15 K.

5.5 The Result of the Simulation

For objective reasons, the competent authority, represented by the Romanian Naval Authority throught Romanian SAR Agency, fails to intervene with ships and depollution equipment to locate, limit spreading and collect the spilled crude oil on the surface of the sea water so the crude oil slick is transported by the resultant of the speed and direction of the current and the wind towards the shoreline according to Fig. 5.1

So, the crude oil slick moves in the direction of the Eforie Nord resort where it pollutes both the beach and the breakwater of tourist port. The crude oil slick reaches the shoreline in eleven hours and thirty-five minutes after the spill occurred.

In seventeen hours, the beach it is polluted over a distance of 1.335 m, with an amount of 18 tons of crude oil. The volume of mixture (crude oil/water) is 2,040 tons, of which 794 m^3 it is crude oil. The maximum thickness of the crude oil slick is 523 mm, and the slick surface is 51,791 m^2, Fig. 5.2.

In Table 5.1 it is shown the characteristics of the crude oil slick during transport, under the action of marine currents and wind. The data are determined every half hour and include: amount of spilled/floating/evaporated/stranded/floating mixture crude oil and the maximum thickness/area of crude oil slick.

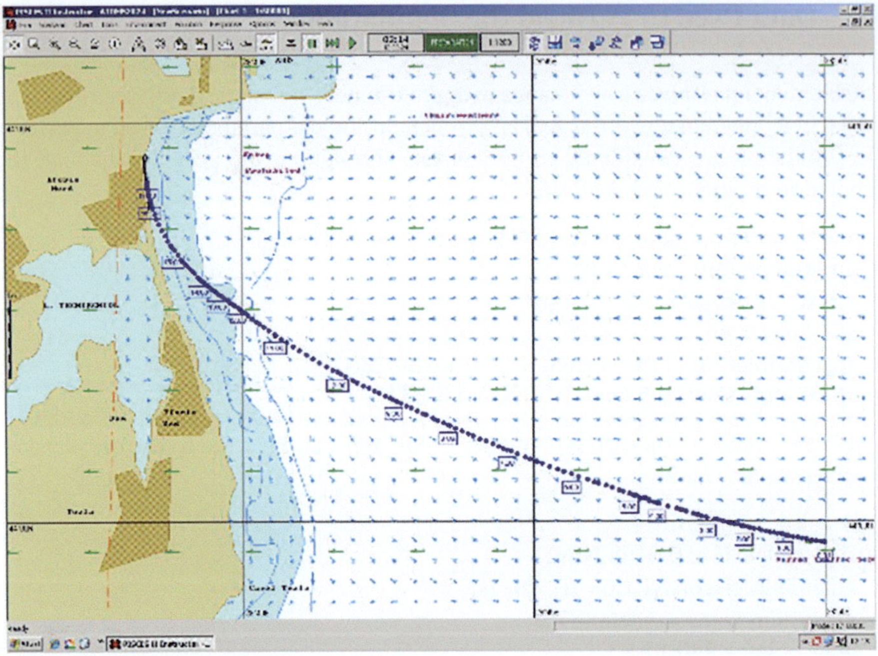

Fig. 5.1 The evolution of oil slick (PISCES29-PL Specifications, ver. 1.0, 2008)

5.6 Conclusions

The EU directive provides strict rules based on the "polluter pays" principle. This means that anyone who causes damage to the environment is responsible for their actions and must take the necessary remedial or preventive measures and bear all related costs to restore the previous situation.

Starting from this EU directive the estimated damages caused to the environment can hardly be counted in money, they are mainly represented by:

- the impact on the entire marine ecosystem (fauna and flora) in entire Eforie Nord resort marine area, an ecosystem that will need decades to recover;
- the impact on the local economy and here we are talking about tourism and fishing industry, industries that will be practically destroyed for many years from now;
- the initial financial impact regarding the effort of the local authorities to clean the beach and the breakwater of tourist port as quickly as possible.

Analyzing the price of a barrel of crude oil, which at the moment is approximately \$ 73/barrel, we can estimate that the accounting value of the amount of crude oil spilled into the sea is approximately \$ 480,048, a small price compared to the harm done to the environment.

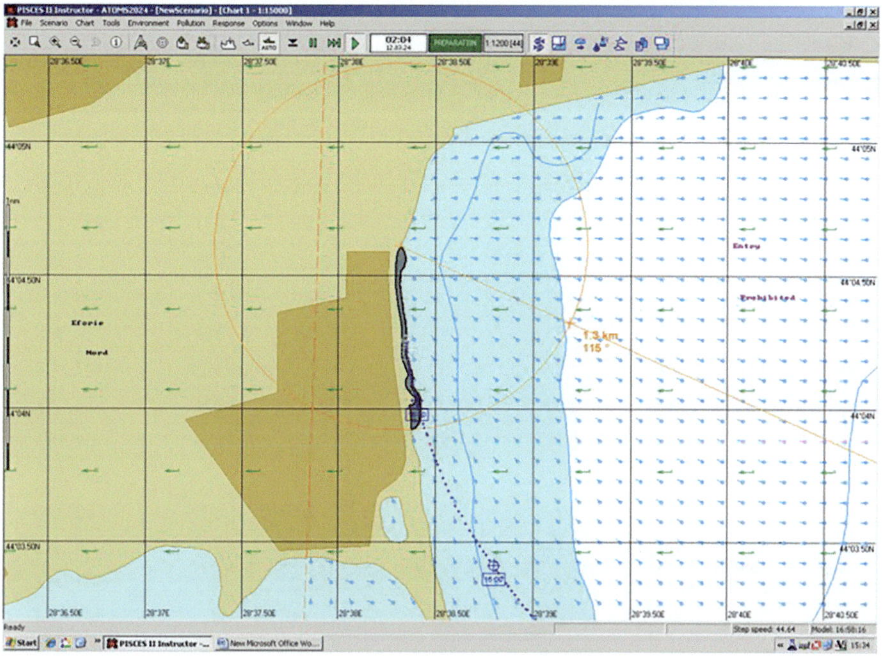

Fig. 5.2 Crude oil slick on the shoreline (PISCES29-PL Specifications, ver. 1.0, 2008)

Considering the consequences that could occur as a result of the failure to organize an effective intervention to locate, limit and collect crude oil spilled on the surface of the sea, the competent authority must test the Intervention Plan for Black Sea with a well-defined frequency, in different scenarios, to train the personel and to check the operative response and intervention times, so as not to reach a situation similar to the one described above.

Table 5.1 Characteristic of the oil slick

Time (h:min)	Amount (t)					Max. thickness (mm)	Slick area (m^2)
	Spilled	Floating	Evaporated	Stranded	Floating mixture		
0:00	0	0	0	0	0	0	0
0:30	31.5	31.4	0.1	0	31.8	5.9	18,762
1:00	128	127	0.8	0	130	9.9	73,794
1:30	253	251	2.5	0	261	9.7	152,897
2:00	368	363	5.1	0	386	8.7	237,900
2:30	475	466	8.7	0	505	7.9	329,368
3:00	573	560	13.1	0	620	12.8	422,803
3:30	666	648	18.3	0	731	9.9	513,717
4:00	754	729	24	0	841	6.8	604,881
4:30	837	807	30.3	0	950	6.9	695,533
5:00	919	881	37.1	0	1,059	10.9	783,520
5:30	920	876	43.9	0	1,090	5.4	804,237
6:00	920	869	50.1	0	1,122	4.5	801,961
6:30	920	864	55.7	0	1,154	4	795,694
7:00	920	859	60.8	0	1,188	4	785,522
7:30	920	854	65.3	0	1,223	4.2	771,039
8:00	920	850	69.4	0	1,259	4.4	758,072
8:30	920	846	73.1	0	1,296	4.6	746,282
9:00	920	843	76.5	0	1,334	4.8	732,464
9:30	920	840	79.7	0	1,373	5	727,764
10:00	920	837	82.6	0	1,414	5	740,379
10:30	920	834	85.4	0	1,455	4.8	782,267
11:00	920	831	88.1	0	1,496	4.7	812,906
11:30	920	828	90.8	0	1,539	4.7	803,400
12:00	920	820	93.4	5.5	1,570	14.3	746,696
12:30	920	818	95.6	6	1,614	173	668,509
13:00	920	815	97.5	6.3	1,659	393	592,971
13:30	920	814	99.2	6.3	1,706	406	521,518
14:00	920	812	101	6.7	1,754	548	450,302
14:30	920	810	102	7.9	1,800	494	377,564
15:00	920	806	103	10.5	1,845	604	290,127
15:30	920	801	103	14.4	1,887	504	184,237
16:00	920	799	104	16.6	1,935	525	109,681
16:30	920	797	104	18	1,985	523	59,800
17:00	920	797	104	18	2,040	523	51,791

Acknowledgements This work would not have been possible without the hydrometeorological data obtained from the "Black sea Floating Offshore Wind" (Blow) project. Finally, I would like to thank the Blow project team who generously shared their time and knowledge with me

References

1. Lazuga, K.: Testing of an oil spill mathematical model contained in Pisces II simulator. Sci. J. Univ. Szczec. **32**(104), 105–109 (2012)
2. Delgado, L., Kumzerova, E., Martynov, M.: Simulation of oil spill behaviour and response operation in Pisces. WIT Trans. Ecol. Environ. **88**, 279–292 (2006)
3. Voicu, I., Dumitrescu, L.G., Panaitescu, F.V, Panaitescu, M.: Studies on the oil spillage near shorline. ModTECH IOP Publishing IOP Conf. Ser.: Mater. Sci. Eng. 227 012135 (2017). https://doi.org/10.1088/1757-899X/227/1/012135
4. Voicu, I., Dumitrescu, L.G., Panaitescu, V.N., Panaitescu, F.V.: The financial implications of a discharge of oil product, inside Constanta Sea Port aquatory. *Ştiintă şi inginerie* (Science and Engineering) 30, Bucharest AGIR Publishing House, pp. 499–508 (2016)
5. PISCES29-PL Specifications, ver. 1.0 (2008)

Chapter 6
Heat Index and Urban Dâmbovita River Temperature Nexus

Cristina Sorana Ionescu, Daniela Elena Gogoase Nistoran, Ioana Opriş, Ştefan Mugur Simionescu, and Cristian Copilău

Abstract The paper examines the characteristics of the heatwaves that occurred in Bucharest City, Romania, during the last three years. Data consisted in hourly air temperature and humidity retrieved from Open Weather and water temperature measured hourly at the exit of Dambovita River from Bucharest with bluetooth loggers. The heatwaves were defined using Matlab and the Thermal Heat Index (THI) was computed. Correlations between hourly air temperature and humidity were carried out for the obtained heatwaves. Temperature and relative humidity corresponding to the time of daily maximum heat index were classified according to threshold values of THI specified by other authors. Furthermore, the study aims to identify the nexus between THI and air temperature with water temperature of Dambovita River. The lag time between air and water daily maxima was identified.

Keywords Heatwave · Thermal heat index · Urban river · Water temperature logger

6.1 Introduction

The combined effect of air temperature and humidity during summers is one of the main drivers that increases the environmental stress in urbanized areas [1]. The two variables that define the thermal heat index (THI), and their changing values can be linked to global warming [2]. Thus, the thermal heat index is considered as a synthetic metrics to assess the effects on citizens' health [3].

C. S. Ionescu · D. E. Gogoase Nistoran (✉) · Ş. M. Simionescu
Department of Hydraulics, Hydraulic Machines and Environmental Engineering, National University of Science and Technology, Politehnica of Bucharest, Bucharest, Romania
e-mail: daniela.nistoran@upb.ro

I. Opriş · C. Copilău
Department of Energy Production and Use National University of Science and Technology, Politehnica of Bucharest, Bucharest, Romania

R. Tamas and C. A. Heiman (eds.), *Selected Papers from AToMS 2024*, Lecture Notes on Multidisciplinary Industrial Engineering,
https://doi.org/10.1007/978-3-032-02452-7_6

Rivers that cross urban areas generally have the potential of cooling the air during daylight [4]. But in hot days, the high air temperatures evaporate the river water, thus contributing to an increased humidity.

Therefore, the paper investigates the possible interdependences between THI and the river water temperature. The study is conducted for the 2021–2023 heat waves that occurred in Bucharest City, capital of Romania, crossed by the Dambovita River.

6.2 Data and Location

Dâmboviţa River crosses Bucharest City (Fig. 6.1). Its natural riverbed was channelized along 17.5 km starting from Morii Reservoir to its exit from the city [5].

To assess the influence of THI on urban water course, temperature and relative humidity data were obtained from Open Weather [6] for the summers of 2021–2023

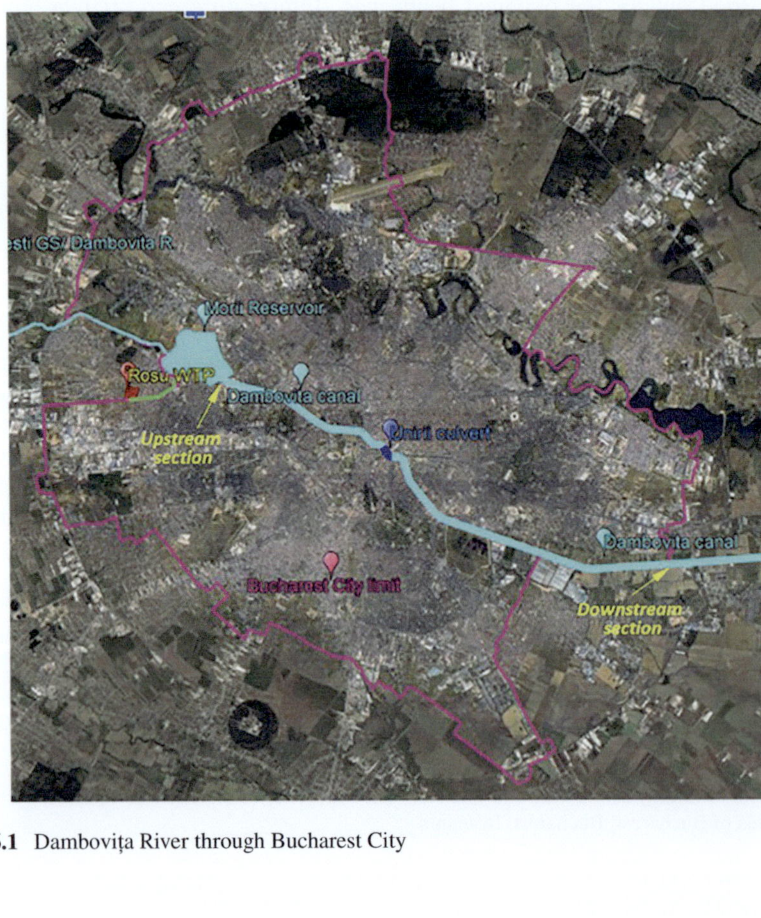

Fig. 6.1 Dambovița River through Bucharest City

periods. Water temperature values were measured during consecutive hottest days of 2021 and 2023 summers using MX2201 temperature sensors equipped with bluetooth communication technology [7], at the exit of Dâmboviţa River from Bucharest City.

6.3 Method

Each country has its particular method for determining heatwaves. Two threshold are used in Romania for defining a heatwave. These are established as being a period of minimum five consecutive days within which the maximum temperature values occurring during the day and the night exceed 35 °C and 20 °C respectively [8].

Matlab was used to identify the heatwaves corresponding to the study period 2021–2023 [9]. For the identified heatwaves, the Thermal Heat Index (THI) was computed in terms of air temperature, T (in °C) and relative humidity, H (in %) with the formula given by Brooke Anderson et al. [10] and Fischer et al. [11]:

$$\begin{aligned} THI = {} & -8.7847 + 1.6114 \cdot T - 0.012308 \cdot T^2 + \\ & + H \cdot \left[2.3385 - 0.14612 \cdot T + \left(2.2117 \cdot 10^{-3}\right) \cdot T^2\right] + \\ & + H^2 \cdot [-0.016425 + \left(7.2546 \cdot 10^{-4}\right) \cdot T + \\ & + \left(-3.582 \cdot 10^{-6}\right) \cdot T^2 \end{aligned} \tag{6.1}$$

To investigate the possible nexus between THI and the temperature of Dambovita River at its exit from Bucharest several correlations between the input data were performed for the identified heatwaves.

6.4 Results and Discussions

Figures 6.2 and 6.3 show the variation with respect to time of air temperature, relative humidity and THI for the 2021 and 2023 identified heatwaves.

Correlations between hourly air temperature, relative humidity and THI were carried out for the 2021–2023 heatwaves (Fig. 6.4). Most of the temperature values range between 22 and 37.5 °C, for a relative humidity interval of 22–84%. However, the maximum air temperature values exceeded 40 °C and the relative humidity dropped towards 20%.

From Figs. 6.5 and 6.6 an average lag time of 4 h can be identified between the moments of peak air temperature and corresponding peak water temperature at the exit of Bucharest City. This is the time needed for the heat transfer from air to water to occur.

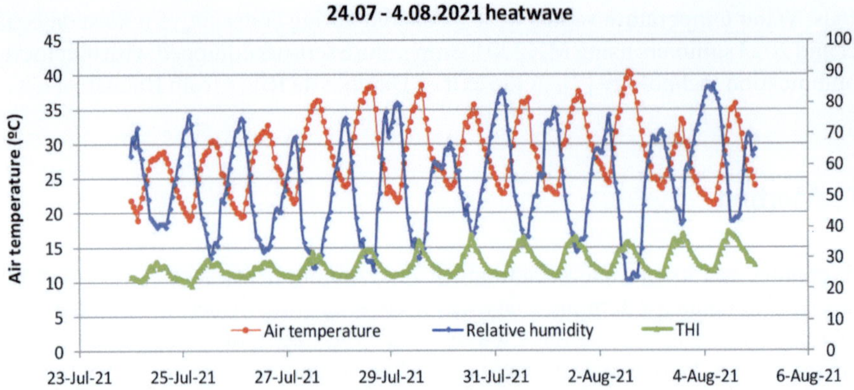

Fig. 6.2 Air temperature, relative humidity and THI during the 2021 heatwave

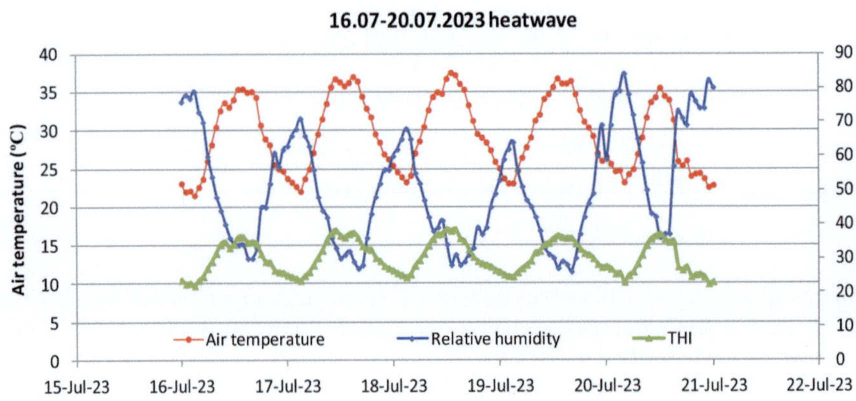

Fig. 6.3 Air temperature, relative humidity and THI during the 2023 heatwave

Fig. 6.4 Air relative humidity—temperature correlation for 2021–2023 heatwaves

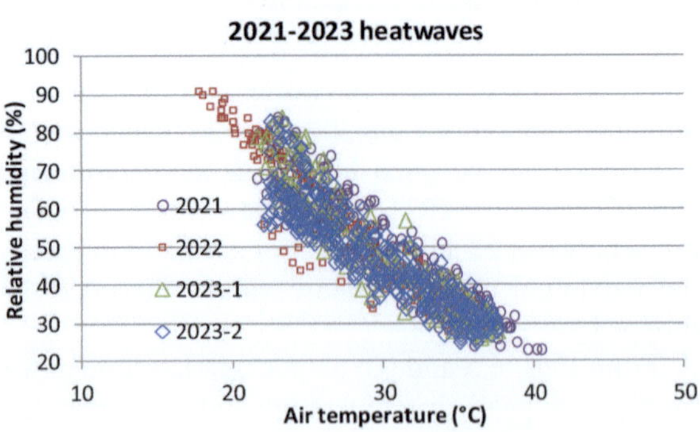

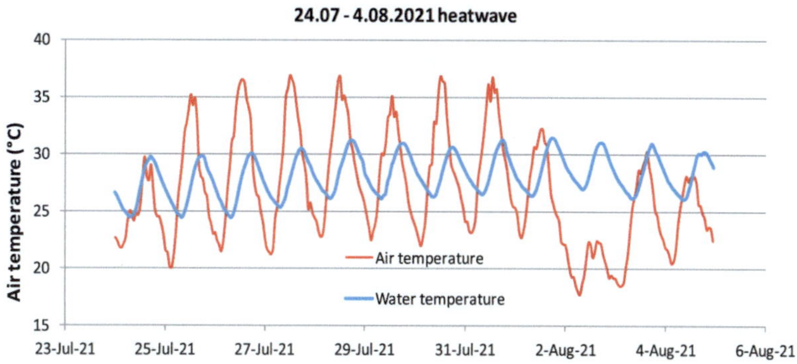

Fig. 6.5 Air and water temperature during the 2021 heatwave

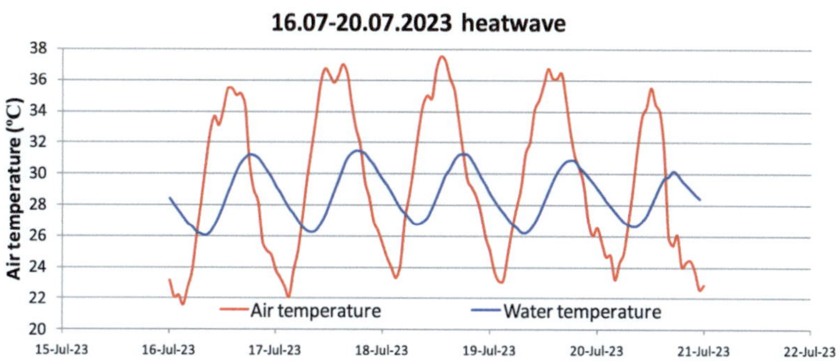

Fig. 6.6 Air and water temperature during the 2023 heatwave

An interesting result was obtained conducting correlations between THI and Dambovita water temperatures at the exit of Bucharest for the two studied heatwaves (Figs. 6.7 and 6.8). The results show a hysteresis effect of water temperature on THI due to the water thermal inertia. between the maximum/minimum THI and the corresponding maximum/minimum water temperature was found. Thus, when THI increases, the water temperature also increases but with aforementioned delay, generating the shape of the lower branch of the graphs in Figs. 6.5 and 6.6. The same stands for the upper branch, when THI decreases and the water temperature also decreases with a similar delay. During the heatwaves, the water temperature varies between 25 and 32 °C.

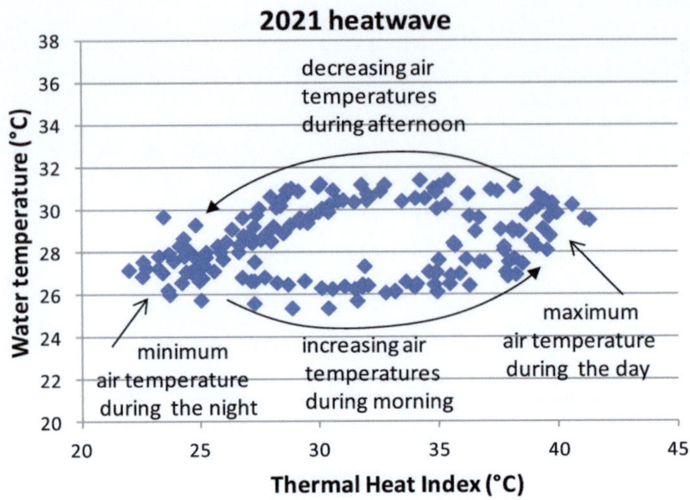

Fig. 6.7 Water temperature dependance on THI during 2021 heatwave

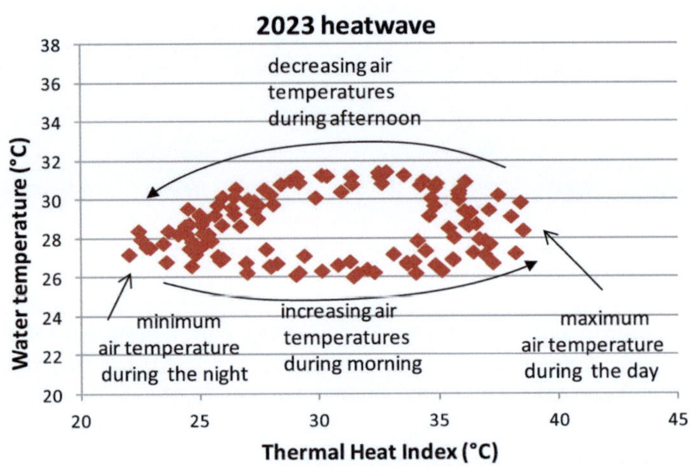

Fig. 6.8 Water temperature dependance on THI during 2023 heatwave

Air temperature and relative humidity corresponding to the moment of daily peak heat index were classified according to threshold values of THI specified by Wehner et al. [12] (Fig. 6.9):

- caution, in the 26–32 °C range
- extreme caution, in the 32–39 °C range, and
- danger, over 39 °C.

Fig. 6.9 Classes of Humidity—Air temperature at the moment of maximum daily THI by thresholds

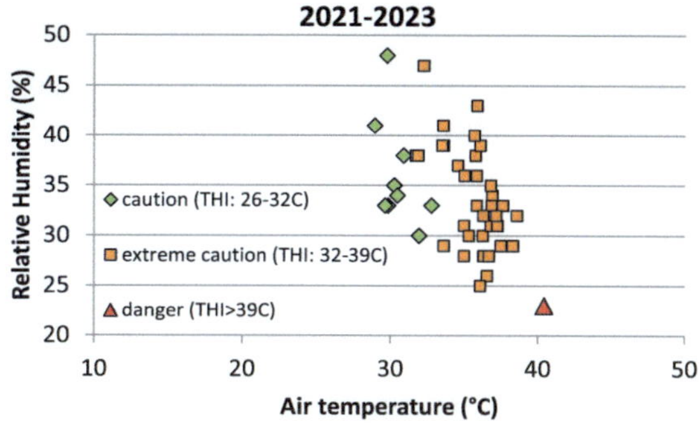

Figure 6.10 depicts the water–air temperature correlation corresponding to the moment of daily peak THI by the same thresholds, showing its influence on the water temperature.

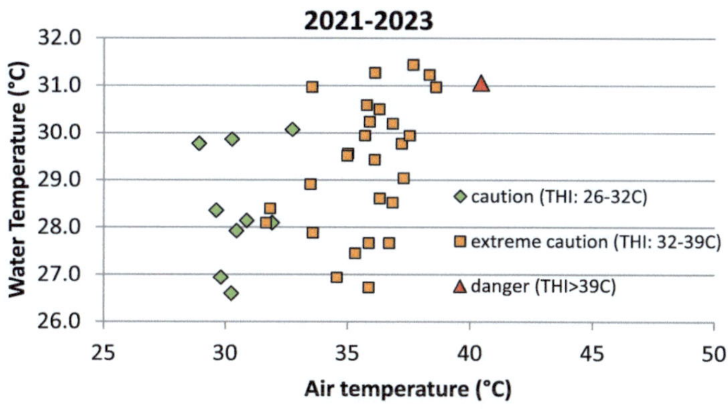

Fig. 6.10 Classes of Water—Air temperature at the moment of maximum daily THI by thresholds

6.5 Conclusions

The results obtained in this study conducted on the urbanized reach of Dambovita River indicate that during summer heatwaves the THI falls mainly into the extreme caution range, inducing threats not only on the population health, but also on the aquatic fauna.

Under increased warming triggered by climate change there is a need to extend the research to include other parameters that influence water temperature and to find sustainable methods to decrease it.

Acknowledgements This work was supported by a grant from the National Program for Research of the National Association of Technical Universities—GNaC ARUT 2023 (no 174/4.12.2023).

Discharge data and access to install the temperature loggers for the present study were kindly provided by the National Water Administration "Romanian Waters".

References

1. Yang, L., Qian, F., Song, D.X., Zheng, K.J.: Research on urban heat-island effect. Procedia Eng. **169**, 11–18 (2016). https://doi.org/10.1016/j.proeng.2016.10.002
2. Romps, D.M.: Heat index extremes increasing several times faster than the air temperature. Environ. Res. Lett. **19**, 041002 (2024)
3. Hoang, T.L.T., Dao, H.N., Cu, P.T., Tran, V.T.T., Tong, T.P., Hoang, S.T., Vuong, V.V., Nguyen, T.N.: Assessing heat index changes in the context of climate change: a case study of Hanoi (Vietnam). Front. Earth Sci. **10**, 897601 (2022). https://doi.org/10.3389/feart.2022.897601
4. Ionescu, C.S., Opriş, I., Gogoaşe Nistoran, D.E., Copilău C.: Simple artificial neural model to predict Dambovita river temperature affected by Urban Heat Islands in Bucharest City. Remote Sens. **16**, 1513 (2024)
5. Stematiu, D., Teodorescu, D.R.: Dâmboviţa in Bucharest–flood defense system. In: Proceedings of the Annual conference of the Romanian Academy of Technical Sciences (in Romanian). Annual conference of the Romanian Academy of Technical Sciences (in Romanian), Bucharest, Romania, 143–150 (2012)
6. OpenWeather. https://openweathermap.org/accuracy-and-quality
7. Onset.: 470 MacArthur Blvd., Bourne, MA 02532, USA (2024). https://www.onsetcomp.com/products/data-loggers/mx2201#specifications. Accessed 21 on March
8. Piticar, A., Cheval, S., Frighenciu, M.: A review of recent studies on heat wave definitions, mechanisms, changes, and impact on mortality, Forum geografic. Studii şi cercetări de geografie şi protecţia mediului, col. XVIII, iss. **2**, 103–220 (2019)
9. Gogoaşe-Nistoran, D.E., Antohe, D.M., Opriş, I., Ionescu, C.S.: A method to analyse urban heatwaves. Case study-Bucharest City (Romania), E3S Web of Conferences 286, 04004 (2021), TE-RE-RD 2021. IOP Conf. Ser.: Earth Environ. Sci. 664 01202652 (2021)
10. Brooke Anderson, G., Bell, M.L., Peng, R.D.: Methods to calculate the heat index as and exposure metric in environmental health research. Environ. Health Perspect. **121**(10), 1111–1119 (2013)
11. Fischer, E.M., Schär, C.: Consistent geographical patterns of changes in high-impact European heatwaves. Nat. Geosci. **3**, 398–403 (2010)
12. Wehner, M., Castillo, F., Stone, D.: The impact of moisture and temperature on human health in heatwaves. Natural Hazard Science, Oxford Research Encyclopedia, (2017). https://doi.org/10.1093/30/9780199389407.013.58

Chapter 7
Preliminary Exploration of Underwater Debris in Lake Como, Italy During the High Tourist Season for Environmental Protection

Jassica Lawrence, Ramona Giurea, Nicola Castelnuovo, and Roberta Bettinetti

Abstract This study presents baseline information on the status of debris accumulation on Lake Como sediments, including the type and quantity of debris over various months during the high tourist season. The study highlights its ecological and human impacts and suggests the management strategies necessary for environmental protection. This research contributes to environmental conservation efforts in Lake Como by encouraging the implementation of measures to prevent lake pollution. The results indicate that in May, 70% of the objects found in Lake Como were made of metal, while 41% were made of plastic. This raise concerns as Lake Como is a frequented spot for tourists and recreational activities. Therefore, understanding the pattern of debris accumulation over time and identifying debris hotspots can be useful in taking proactive measures to protect the environment.

Keywords Environmental protection · Management strategies · Debris · Impacts · Pollution prevention · Tourist attraction

J. Lawrence (✉)
Department of Science and High Technology, University of Insubria, Como, Italy
e-mail: jlawrence@uninsubria.it

R. Giurea
Department of Industrial Engineering and Management, Lucian Blaga University of Sibiu, Sibiu, Romania
e-mail: ramona.giurea@ulbsibiu.ro

N. Castelnuovo
Proteus Assiciation, Como, Italy
e-mail: castelnuovonicola@hotmail.com

R. Bettinetti
Department of Human Sciences, Innovation and Territory, University of Insubria, Como, Italy
e-mail: roberta.bettinetti@uninsubria.it

R. Tamas and C. A. Heiman (eds.), *Selected Papers from AToMS 2024*, Lecture Notes on Multidisciplinary Industrial Engineering,
https://doi.org/10.1007/978-3-032-02452-7_7

7.1 Introduction

Lake Como is a famous tourist destination particularly Como Bay highest number of tourists arrive in Spring and Fall with a weekly rise on weekends. There is an average upsurge of 14.0% during weekends and 10.8% in peak tourism period in high season weekends than weekdays in the low season, thus suggesting patterns are strongly affected by recreational and tourism activities on the lake [1]. The major municipalities are Como and Cernobbio. Their populations, 84,250 and 6,498 inhabitants, give a density of 2,300 and 536 inh./km^2, respectively as reported by ISTAT [2–6, 6]. Potable water consumption, wastewater, municipal solid waste, are related and also influenced by the tourism of people living particulary in NW Lombardy and neighbouring Canton Ticino (Switzerland).. Lake Como is a prominent famous place for tourism, culture, and historical prose heritage. It serves multipurpose i.e. aesthetics, water sports, recreational, irrigation hydropower, and flood risk prevention alongside shores. Lake Como is an important lake in the North of Italy. It also collects water utilized for irrigation in the Po Valley downstream [7].

Freshwater lakes play an essential role in rendering ecosystem services [8]. Anthological activities create disturbance to the core of the ecosystem. Thus, it harms key aquatic organisms such as phytoplankton, fish, and most importantly zooplankton are impacted by lake pollution [9]. Underwater monitoring in lakes provides useful information on lake pollution particularly the debris objects being accumulated on the sediments due to human activities and impacts the health of lakes, rivers, oceans, and the ecosystem [10, 11].

Underwater debris is human-produced waste found in water bodies and is a serious environmental concern. The debris puts enormous risk to aquatic ecosystems, human health, and water transportation. Such as debris damaged the habitat of 4 to 10 million crabs a year in Louisiana [1] and caused damages like propeller entanglement to fishing boats [12]. Therefore, it is vital to monitor the debris emergence and warn the authorities to take preventive measures to avoid possible risks [13]. The purpose of this study is to indicate the predominant objects and materials found underwater on lake sediments, particularly during summer when tourism is at its peak suggested mitigation measures for environmental protection and help environmental management authorities to take better management steps to prevent environmental damage.

7.2 Methods

Study area: The study site Village, Lake Como is a peninsula Villa Geno locality, extending slightly to the centre of the first part of the Lake Como basin (Lake Como, Northern Italy), 1 km north of the city of Como.

Tool: The drone Remote Operating Vehicle (ROV) was used to record the debris objects. It is a water drone, attributed to reasonable price, are optimum devices to record underwater videotaping. The model used in the present is a Trident from Open Rov snc. Trident is associated with Raspberry Pi shield programmed to record a video from a CCD sensor and to manage information from a controller, cabled to the surface with a 100 m tether. It is assembled with LED lights (approximately 200 luminosity of brightness), thermometer, depth sensor and internal digital compass for navigation. In this study, a smart phone (iPhone) was used to control the drone and application.

Littoral videos in the months of (April, May, June, August, and November 2019) were considered to count the number of objects due to better quality in the same area to understand the pattern of accumulation, The drone Remote Operating Vehicle (ROV) was installed 10 m depth below the water surface.

Video Quality: Each month has six video transects to understand and count the debris. For this study, it was better to consider littoral HQ videos rather than profound ones due to the better quality.

7.3 Results and Discussion

Among the debris, metal, and plastic objects were the predominant material types found in May, while glass and wood debris were the least common. In May the highest count of metal objects was, i.e. 65, followed by 43 in August and 39 in November. While 31 in June and the lowest 25 were counted in April (Figs. 7.1 and 7.2).

In Fig. 7.3 among the plastic objects the highest number of total plastic objects, were in April i.e. 43. The dominant plastic object found in all months was candy wrapping packaging. The highest candy wrapping packaging counted was in April and November i.e. 20, followed by May 15, August 13 and June 6. The second highest object found in all months was disposable cup particularly highest in April i.e. 8, followed by November 7, May 4, August 2 and June 1. The rest of the objects were not present in all the months constantly. The highest count was attributed to the fact that April is one of the peak months for tourism, although the entire season is full of tourists due to school holidays and favourable weather conditions [2]. It has been concluded based on preliminary monitoring and quantification, that there is potential waste accumulated at the bottom of the lake. Therefore, it is important to understand the pattern and sources of this waste to effectively mitigate and manage the underwater debris.

The most common plastic object was candy wrapping packaging found in all the months. The highest counted was in April and November i.e. 20. The second highest candy wrapping packaging was counted in May i.e. 15. Besides, in April, 8 disposable cups, 6 short plastic pipes, 4 plastic bottles, 3 plastic chairs, 1 plastic shovel, and 1 table were counted. In May, 7 plastic chairs, 4 disposable cups, 4 plastic bottles, 2 short plastic pipes, 2 tires, 1 plastic shovel, 1 plastic table, 1 croc shoe, 1 disposable fork, 1 plastic wire, and 1 plastic toy were counted.

In June, 6 candy wrapping packaging, 2 tires, 2 umbrella plastic parts, 1 disposable cup, 1 plastic chair, 1 short plastic pipe, 1 toy, 1 lamp stand, 1 camera stand, and

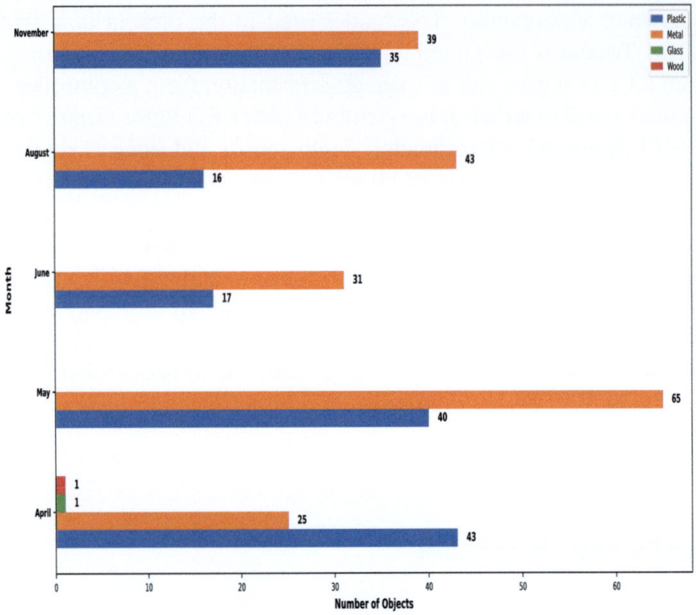

Fig. 7.1 Material category distribution with respect to months

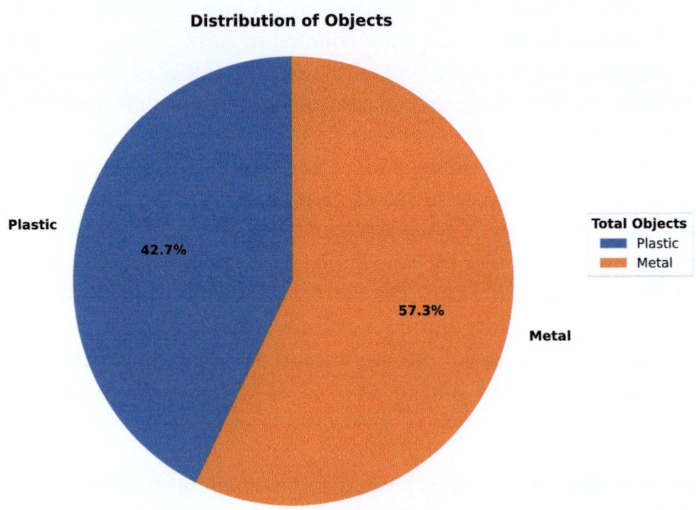

Fig. 7.2 Debris material composition by percentage

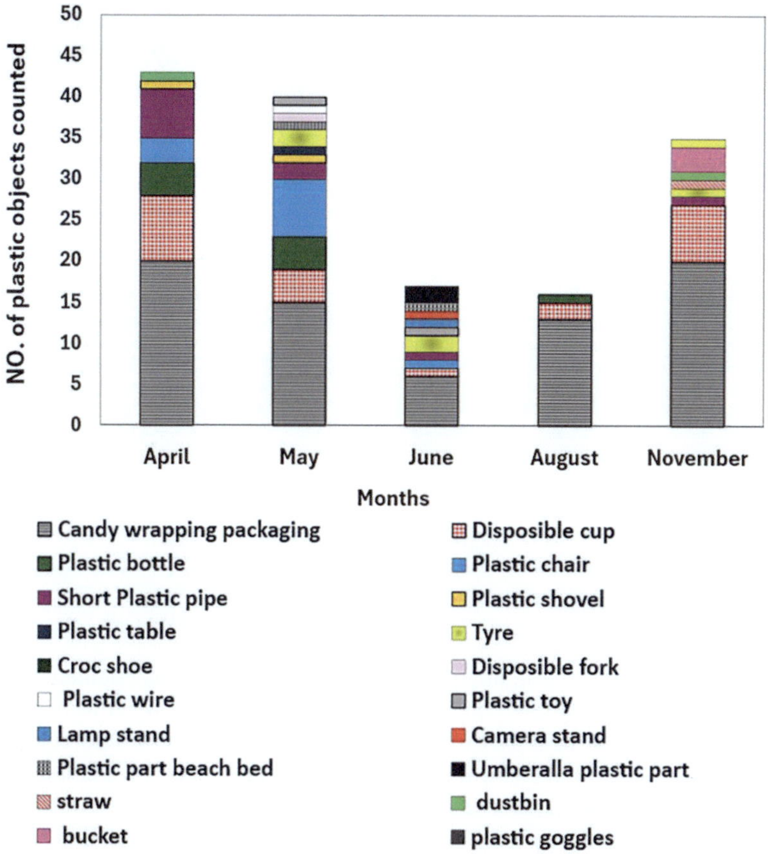

Fig. 7.3 Total plastic objects per month

1 plastic part beach bed were counted. Whereas, in August 13 candy wrapping packaging, 2 disposable cups, and 1 plastic bottle were counted. Finally, in November, 7 disposable cups, 3 buckets, 1 short plastic pipe,1 tire,1 straw, 1 dust bin, and 1 plastic goggles were counted. The results show that metal and plastic were present on the Lake Como sediments. Anthropogenic activities are the fundamental causes of eutrophication and water pollution (Figs. 7.4 and 7.5).

Among the metal objects, the highest count was observed in May, i.e. 65. The dominant metal object found in all the months was soda tin. The highest count of soda tin was in the month of May i.e. 48 followed by August, 37 and November, 30. The second most object found in all the months was the beach bed. The highest count was observed in April (9) followed by May (8) and November (7), June (5) and the least was in August (3 were counted). The rest of the objects were not present in all the months constantly.

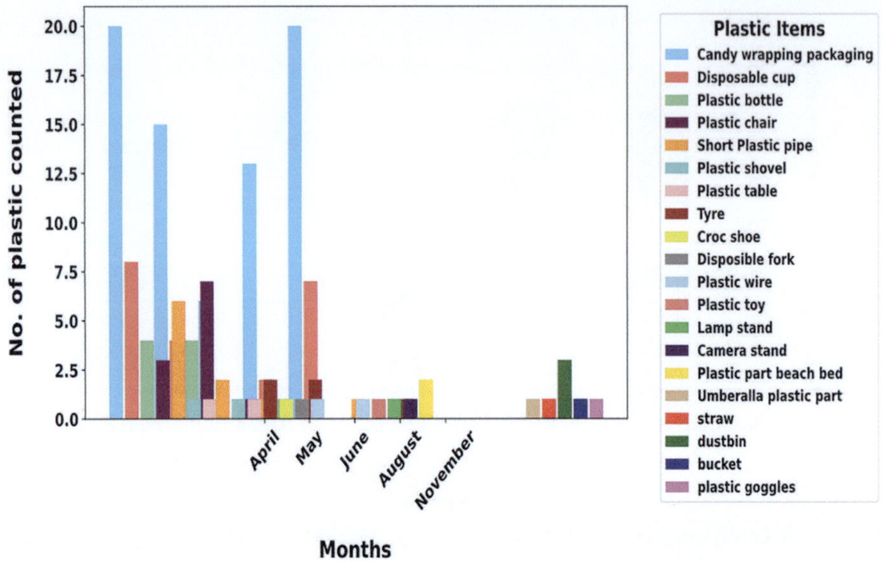

Fig. 7.4 Total plastic objects per month

In Fig. 7.6 the object table in April, May and November was counted 1 of each. Whereas in April iron cylinder, metal piece of boat, and metal sheet were counted 1 of each. While in May metal star object, iron stand, metal part of umbrella and iron bed were counted to be 1 of each. While in June, 2 iron cabinets and 2 umbrellas were counted. Also, metal wire, lamp stand, and camera stand 1 of each were counted. While in August, 2 iron stands and 1 iron basket were counted. Finally, in November the other counted object was iron cylinder i.e..1.

In this study Lake Como concerning seasonal and tourist aspects was investigated from May 2020 to August 2021. The intense pressure was found on Lake Como particularly during weekends in summer (up to +14% than weekdays in winter), indicating tourism is the crucial reason. The results suggested a strong correlation between human activities and lake-water quality related to human activities and tourism [1].

Several studies investigated the impacts of water pollution and tourism ventures. The West Lake Basin in China was investigated to understand the impacts of tourism and pollution, on touristic business income and tourist waste spike and suggested that these variables have a direct correlation with each other [14]. Similarly, in our study, the debris objects were certainly present in all the tourist seasons on Lake sediments.

Ecological and human implications of debris accumulation: Debris accumulation by humans threatens aquatic organisms leading to detrimental ecological effects. Almost 82% of impacts were caused by plastic relative to other materials (e.g., metals, glass) and largely (89%) at sub-organismal levels (e.g., molecular, cellular, tissue) of aquatic organisms. Also, the other impacts were caused majorly by plastic

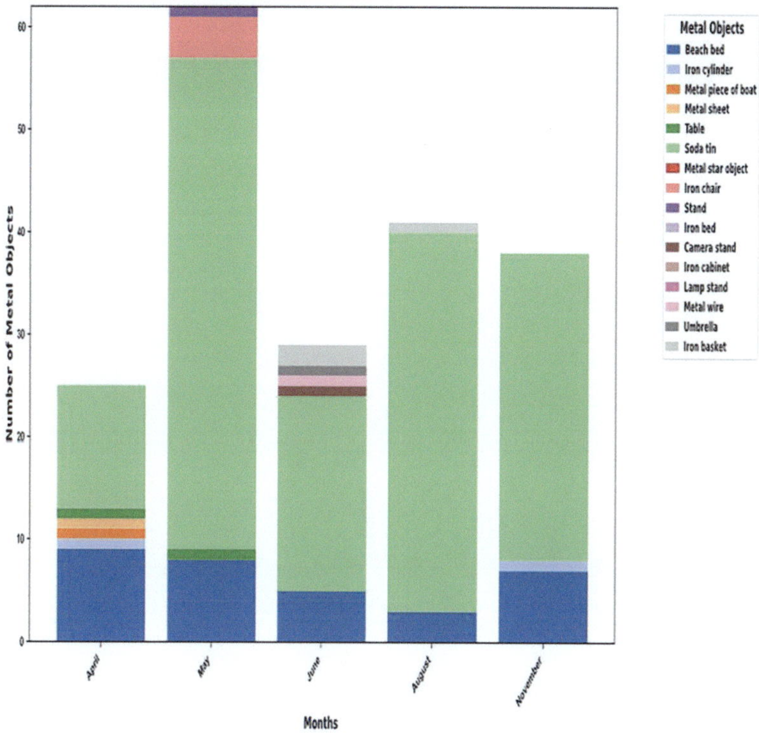

Fig. 7.6 Total count of monthly metal objects per month

debris objects such as rope, straws, and fragments leading to accelerated mortality of organisms, and disturbing the entire food web This study and all the literature privies sufficient evidence of debris accumulation in Lake Como sediments to take mitigation measures by the management and policymakers to prevent environmental pressure to go beyond repair [15].

7.4 Conclusions

The waste objects accumulate on lake sediments. However, the objects such as chairs and beach beds constantly accumulate due to nearby commercial restaurants. Due to lake sedimentation the objects underwater get covered in sand and some of them are moved due to tides. This unusual dynamic needs more research in detail given its serious impact. However, this study can be a guide to highlight the issue in Lake Como and stimulate the environmental agencies to focus on sustainable tourism and pollution prevention.

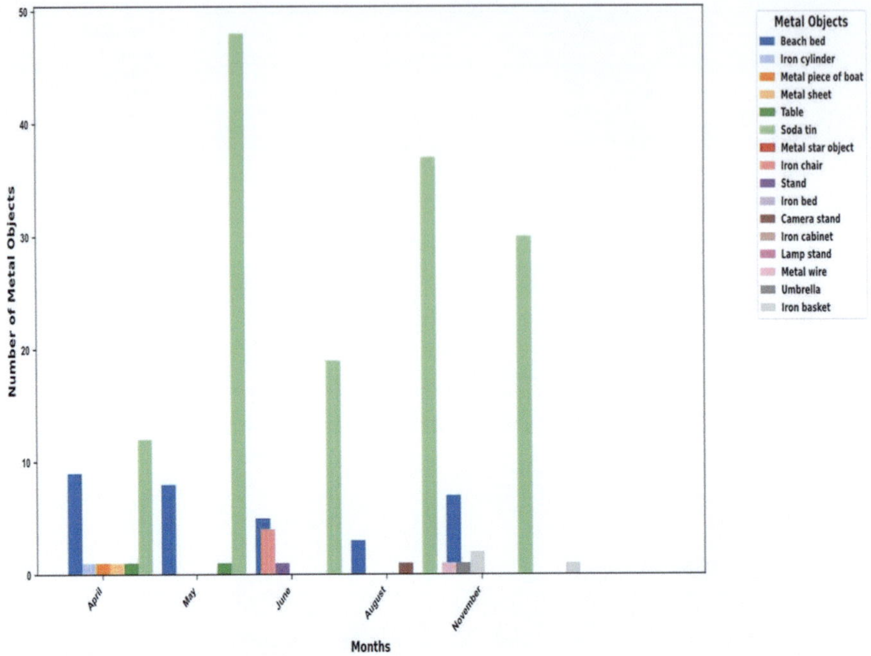

Fig. 7.6 Comparison of metal objects per month

References

1. Vavassori, O.D., Brovelli, M.A.: Population space–time patterns analysis and anthropic pressure assessment of the Insubric Lakes using user-generated geodata. ISPRS Int. J. Geo Inf. **11**(3), 206 (2022). https://doi.org/10.3390/ijgi11030206
2. Giussani, D.C., Monticelli, D., Pozzi, A., Recchia, S.: A chemometric approach to the investigation of major and minor ion chemistry in Lake Como (Lombardia, Northern Italy). Anal. Chim. **96**(5–6), 339–346 (2006). https://doi.org/10.1002/adic.200690035
3. Ragazzi, M., Conti, F., Torretta, V., Romagnoli, F., Zatelli, C., Ghiringhelli, G., Lakatos, E.S., Rada, E.C.: MSW management in two Italian mountainous areas. Environ. Clim. Technol. **28**(1), 84–93 (2024). https://doi.org/10.2478/rtuect-2024-0008. Accessed 8 Jul 2024
4. Rada, E.C., Zatelli, C., Mattolin, P.: Municipal solid waste selective collection and tourism. WIT Trans. Ecol. Environ. **180**, 187–197 (2014)
5. Lichtmannegger, T., Hell, M., Wehner, M., Ebner, C., Bockreis, A.: Seasonal tourism's impact on wastewater composition: evaluating the potential of alternating activated adsorption in primary treatment. Sci. Total Environ. **926**, 171869 (2024). https://doi.org/10.1016/j.scitotenv.2024.171869
6. Schiavon, M., Giurea, R., Ionescu, G., Magaril, E., Rada, E.C.: Agro-tourism structures, SARS-CoV-2: the role of water. MATEC Web Conf. **354**, 00070 (2022). https://doi.org/10.1051/matecconf/202235400070
7. Fuso, F., Casale, F., Giudici, F., Bocchiola, D.: Future hydrology of the cryospheric driven Lake Como catchment in Italy under climate change scenarios. Climate **9**(1), 8 (2021). https://doi.org/10.3390/cli9010008

8. Dodds, W.K., Perkin, J.S., Gerken, J.E.: Human impact on freshwater ecosystem services: a global perspective. Environ. Sci. Technol. **47**(16), 9061–9068 (2013). https://doi.org/10.1021/es4021052
9. Bettinetti, R., Quadroni, S., Boggio, E., Galassi, S.: Recent DDT and PCB contamination in the sediment and biota of the Como Bay (Lake Como, Italy). Sci. Total Environ. **542**, 404–410 (2016). https://doi.org/10.1016/j.scitotenv.2015.10.099
10. Statzner, B.P., Dolédec, S., Schöll, F.: Invertebrate traits for the biomonitoring of large European rivers: an initial assessment of trait patterns in least impacted river reaches. Freshw. Biol. **50**(12), 2136–2161 (2005). https://doi.org/10.1111/j.1365-2427.2005.01447.x
11. Fornaroli, R., Cabrini, R., Zaupa, S., Bettinetti, R., Ciampittiello, M., Boggero, A.: Quantile regression analysis as a predictive tool for lake macroinvertebrate biodiversity. Ecol. Ind. **61**, 728–738 (2015). https://doi.org/10.1016/j.ecolind.2015.10.024
12. California state water resources.: Problem with marine debris. http://waterboards.ca.gov/waterissues. Cerca con Google. https://www.google.com/search?client=firefox-b-d&q=Problem+with+marine+debris.+http%3A%2F%2Fwaterboards.ca.gov%2Fwater+issues (2002). Accessed 26 Mar 2024
13. Wang, Y., Tan, R., Xing, G., Wang, J., Tan, X., Liu, X., Chang, X.: Aquatic debris monitoring using smartphone-based robotic sensors. In: IPSN-14 Proceedings of the 13th International Symposium on Information Processing in Sensor Networks. Berlin, 13–24 (2014). https://doi.org/10.1109/IPSN.2014.6846737
14. Sun, Q., Liu, Z.: Impact of tourism activities on water pollution in the West Lake Basin (Hangzhou, China). Open Geosci. **12**(1), 1302–1308 (2020). https://doi.org/10.1515/geo-2020-0119
15. Markogianni, V., Mentzafou, A., Dimitriou, E.: Assessing the impacts of human activities and soil erosion on the water quality of Plastira mountainous Mediterranean Lake, Greece. Environ. Earth Sci. **75**(10), 915 (2016). https://doi.org/10.1007/s12665-016-5737-8

Part II
Materials, Biomaterials, and Biological Applications

Chapter 8
Study of the Calcium Phosphate Glasses Doped with Silver and Copper I. Synthesis and Characterization of the Glasses

Daniela Avram, Dan Nicolae Ungureanu, Ionica Ionita, Ana-Maria Hossu, and Cosmina Cristiana Avram

Abstract Calcium phosphate-based biomaterials have been the subject of most preclinical studies to obtain compounds compatible with the human body in terms of bone reconstruction. The most used phosphocalcic materials as bone substitutes are calcium phosphate ceramics, bioglasses and glass–ceramics, calcium phosphate cements, phosphate coatings and calcium phosphate composites.

Keywords Biomaterials · Bioglasses · Sol–gel method · Copper · Silver

8.1 Introduction

The introduction of silver and copper into the structure of bioactive glass and other materials with medical uses is a relatively recent technique for obtaining biomaterials for orthopedic and maxillofacial reconstruction and aims to reduce the risk of bacterial contamination of the implant [1]. Recent studies have shown that bioactive glass obtained by the sol–gel technique doped with monovalent copper ions has a high bactericidal potential and, in addition, exhibits anti-inflammatory properties [2]. The introduction of silver, copper and other metals with bactericidal potential into

D. Avram (✉)
Department of Food Engineering, Valahia University of Targoviste, Targoviste, Romania
e-mail: daniela.avram@valahia.ro

D. N. Ungureanu
Department of Material Engineering, Valahia University of Targoviste, Targoviste, Romania

I. Ionita · A.-M. Hossu
Department of Sciences and Advanced Technologies, Valahia University of Targoviste, Targoviste, Romania

C. C. Avram
SC Almina Trading SA, Targoviste, Romania

R. Tamas and C. A. Heiman (eds.), *Selected Papers from AToMS 2024*,
Lecture Notes on Multidisciplinary Industrial Engineering,
https://doi.org/10.1007/978-3-032-02452-7_8

the composition of biomaterials has been preferred to the introduction of antibiotics (gentamicin, vancomycin, cephalexin, etc.) [3–7] because most strains of microorganisms have become resistant to drugs, and their effectiveness in preventing and combating infections has become limited, not to mention the extremely high costs in their use [8]. Russel and Hugo studied the antimicrobial activity of silver and copper [9], and Marsh of zinc [10]. They reach the unanimous conclusion that only the ionic forms of the respective metals have an antibacterial effect. Calcium phosphate glasses doped with copper by the sol–gel technique are partially soluble materials that can act as a unique system for the controlled release of copper ions [11, 12].

The ions are embedded in the glass structure and are not a separate phase; thus, their release rate is predominantly defined by the overall glass biodegradation rate. Since the literature in our country and abroad is quite deficient in information about phosphocalcic materials doped with copper ions, regardless of their oxidation state, 1+ or 2+, we chose to study the properties of calcium phosphate glasses doped with cuprous ions (1+).

8.2 Materials and Method

The sol–gel method for the synthesis of glasses with bioactive potential involves going through some typical framework steps; these consist of catalytic hydrolysis and condensation of oxidic precursors in the form of soil, gelation of the soil in a saturated atmosphere, maturation (ageing) of the gel, its drying and thermal stabilization of the resulting xerogel. The choice of the sol–gel method for the synthesis of the glasses samples under study was mainly because vitreous materials with a predetermined structure can be obtained, depending on the working parameters and the precursors used, which are quite difficult to obtain by conventional techniques for melting oxide mixtures [13]. The improvement and predictability of the properties of the final products makes sol–gel synthesis particularly advantageous and attractive for the field of biocompatible materials compared to the classical method.

Last but not least, this synthesis technique is more feasible because it allows obtaining much lower production costs, keeping properties such as specific surface area and porosity at high values, and the possibility of doping with metals (silver, magnesium, zinc, copper, etc.) that can confer additional special properties (anti-inflammatory, antimicrobial, healing, etc.) without losing their bioactivity. In the present study, the synthesis of three compositions of ternary and quaternary calcium phosphate glasses with the composition shown in Table 8.1 is presented. For the synthesis of ternary glass (PA1 undoped), tetraethylorthosilicate (TEOS)–$Si(OC_2H_5)_4$ for SiO_2 and triethyl phosphate (TEP)–$PO(C_2H_5O)_3$ for P_2O_5 were used as organic precursors, as network formers, and as inorganic precursor for CaO, calcium nitrate tetrahydrate crystallized with 4 water molecules, $Ca(NO_3)_2·4H_2O$ For the doped glasses, $AgNO_3$ was used as precursor for Ag_2O, and for cuprous ion doping, CuCl was used. The composition of the glass doped with cuprous ions PC1 is compositionally equivalent to the glass doped with silver PA1, both compositions are derived from the ternary glass P1, from which 3% SiO_2 has been substituted

Table 8.1 Chemical composition in oxides of the glass (% weight)

Sample code	SiO_2	CaO	P_2O_5	Ag_2O	Cu_2O
P1	50	45	5		
PA1	47*	45	5	3	
PC1	47*	45	5		3

* Indicates which oxides were substituted from the initial composition to introduce the oxides of the dopant materials

in order to introduce the oxides of the two dopant metals, keeping most of the phosphocalcic composition. The framework technology for the synthesis of silver and copper doped quaternary glasses basically goes through the same steps specific to the sol–gel method, i.e. catalytic hydrolysis and condensation of the oxide precursors in order to obtain colloidal sol with a continuous structure, gelation of the sol, gel maturation, gel drying and thermal stabilisation of the resulting xerogel.

8.3 Results and Discussion

The synthesized glass samples were subjected to dispersive X-ray fluorescence spectroscopy by wavelength according to the standard working protocol for the ARL Advant X spectrometer. The results of the WD-XRF analyses of the non-immersed calcium phosphate glasses in simulated body fluid (SBF) are presented below in the form of histograms to which tables of theoretical and practical compositions are attached.

8.3.1 Determination of the Chemical Composition of Calcium Phosphate Glasses Synthesized by Sol–gel Technique Through WD-XRF

Figure 8.1 shows a comparison of the theoretical and real compositions of the three synthesized samples: P1 (a), PA1 (b), PC1 (c). The results of the WD-XRF analysis for the P1 glass (Fig. 8.1a) indicate values of the practical composition very close to those calculated theoretically, with differences of less than 2.3% for SiO_2 and less than 1% for the other two oxides.

These show that the synthesis proceeded under optimal conditions and the working parameters were correctly chosen. Figure 8.1b shows the histogram of the composition of the PA1 silver ion doped calcium phosphate glass (Fig. 8.1b).

Figure 8.1b shows the actual oxide composition of the PA1 glass 48.14% SiO_2, 41.6% CaO, 5.95% P_2O_5 and 3.7% Ag_2O, which corresponds to a very good synthesis

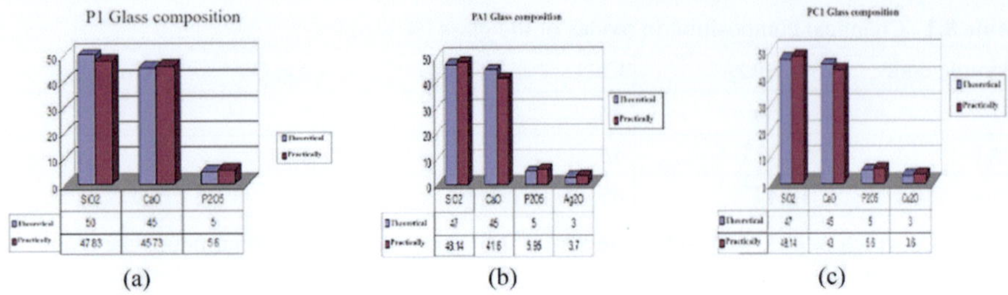

Fig. 8.1 Histograms of P1 (**a**), PA1 (**b**) and PC1 (**c**) glasses compositions

yield of 97.5%. The CaO fixation in the vitreous network is 3.4% lower than calculated, making the values of the other component oxides slightly higher than calculated (up to 1.14% for SiO_2).

Figure 8.1c shows the elemental chemical analysis result for the quaternary calcium phosphate glass doped with cuprous ions, PC1. It is observed that the composition of PC1 glass is 48.14% SiO_2, 43% CaO, 5.6% P_2O_5 and 3.6% Cu_2O. The synthesis yield is 100%.

8.3.2 *Dimensional Analysis and Granulometric Distribution of the Calcium Phosphate Glasses Synthesized by the Sol–gel Method Using Dynamic Pulsed Laser Light Scattering (DLS)*

The dimensional analysis of pulverulent calcium phosphate glasses synthesized by the sol–gel technique is necessary because there is a direct correlation between the size of the glass particles and their degree of biocompatibility, as well as regarding their in vivo behavior in terms of the degree of resorption, therefore implicitly in the way they intervene in bone reconstruction. The glass samples synthesized by the sol–gel technique were subjected to DLS analysis after being completely dispersed by repeated ultrasonication and suspended in sodium acid phosphate–NaH_2PO_4—with variable pH (3–5.5). The result of the DLS analysis performed on the ternary calcium phosphate glass sample, P1 is centralized in Table 8.2. The hydrodynamic diameter (d_H), equivalent to the effective diameter (d_{eff}), was calculated from the autocorrelation function of the light intensity scattered by the particles, assuming that they have a spherical shape.

Table 8.2 The hydrodynamic diameters of ternary glass

Sample code	$D_{Hydrodynamic}$ ($D_{Effective}$) [nm]	Number of populations	D_{Medium} [nm]
P1	716,6 ± 12,7	2	647,9

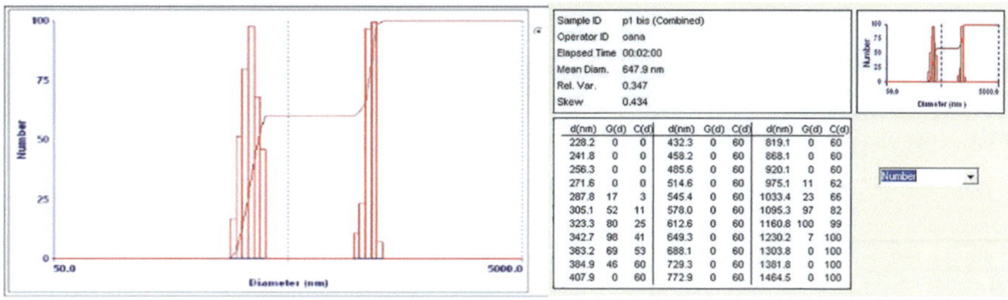

Fig. 8.2 Histograms of mean glass diameter P1

From Table 8.2 the glass sample presents a variable number of populations (families) of particle sizes, from one to three. To present how to interpret DLS analysis data, Fig. 8.2 shows the histogram of the mean diameter calculated as a function of the number of particles dispersed in solution for sample P1.

The results of the DLS analysis performed on the quaternary calcium phosphate glass doped with silver ions, respectively the samples marked PA1 are centralized in Table 8.3. Figure 8.3 shows the histogram of the average diameter calculated according to the number of particles dispersed in the solution for sample PA1. The data in Table 8.3 show that the silver-doped calcium phosphate glass exhibits unimodal particle populations.

The copper ion doped quaternary calcium phosphate glass sample PC1 has the DLS analysis result shown in Table 8.4. The histogram of the copper doped glass sample is shown in Fig. 8.4.

The conclusion of the DLS analysis of the samples of calcium phosphate glasses synthesized by the sol–gel technique is that all three are biocompatible, being able to reach micron and even nanometric sizes through post-synthesis processing. Their capacity for bioactivity and bioresorption is confirmed by XRF, FTIR and SEM

Table 8.3 The hydrodynamic diameters of quaternary glass doped with silver PA1

Sample code	$D_{Hydrodynamic}$ ($D_{Effective}$) [nm]	Number of populations	D_{Medium} [nm]
PA1	527,9 ± 11,38	1	381,4

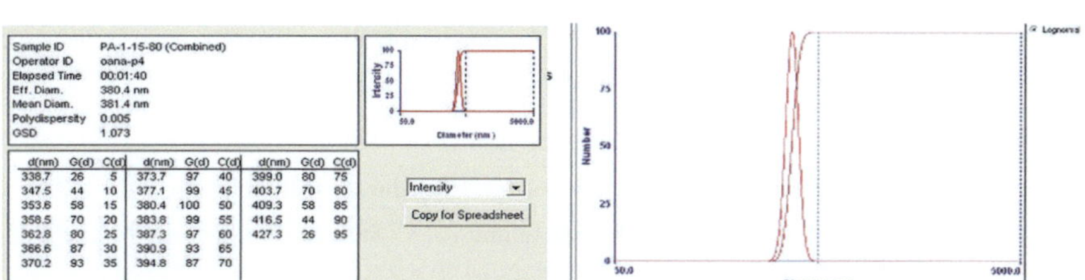

Fig. 8.3 Histogram of the average diameter of PA1 glass

Table 8.4 The hydrodynamic diameters of quaternary glasses doped with copper

Sample code	$D_{Hydrodynamic}$ ($D_{Effective}$) [nm]	Number of populations	D_{Medium} [nm]
PC1	498.6 ± 84.7	1	296.3

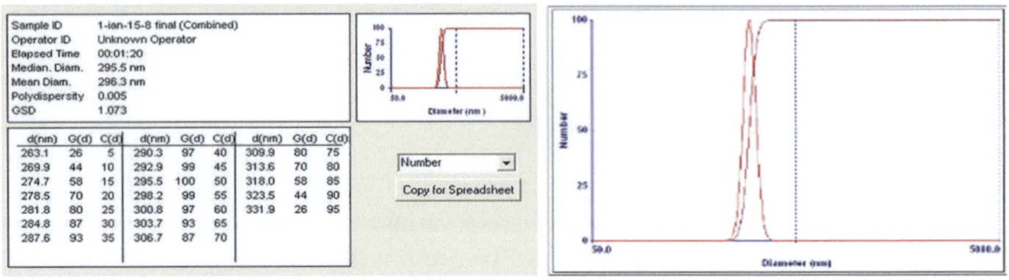

Fig. 8.4 Histogram of the average diameter of PC1 glass

analyzes presented in the second part of this article and in other references in the bibliographic notes.

8.3.3 *Determination of Relative Density by Pycnometric Method*

The relative density of the glasses synthesized by the sol–gel technique was determined by the pycnometric method because calcium phosphate glasses are pulverulent materials and lend themselves to introduction into pycnometers by free flow. The results obtained after determining the density for the synthesized glass samples are presented in Table 8.5.

The density of the PA1 sample doped by 3% Ag_2O indicates that the silver ions from the vitreous mass impart a relatively linear behavior to this glass in terms of properties, as is clear from the DLS analysis which present that they have particles with close average particle sizes. Comparing the density of the ternary P1 glasses with the silver PA1 and copper PC1 doped glasses, respectively, shows that both doped glasses have higher densities than their equivalent undoped composition. Therefore, it can be concluded that the introduction of metals into the vitreous matrix leads to increased density of the glasses. All glass compositions synthesized by the sol–gel

Table 8.5 The relative density of glass powders obtained by the sol–gel method

Sample	Density [g/cm^3]	Sample	Density [g/cm^3]	Sample	Density [g/cm^3]
P1	0,6514	PA1	0,7609	PC1	0,6931

technique have sub-unitary densities. The glass with 3% Cu_2O has a intermediate density between the undoped and silver-doped glass–PA1.

8.4 Conclusions

The sol–gel technique has been successfully used to obtain simple (ternary) and biocide metal (silver and copper) doped calcium phosphate glasses. It proved to be reproducible and controllable when making changes in composition and working parameters. The variation of the stoichiometric compositions of the glasses synthesized by the sol–gel technique was achieved by partial substitution of SiO_2 for silver or copper doping. This resulted in obtaining some glasses with secondary phases in the vitreous structure without negative effects on the bioperformance of the materials.

The copper and silver ions have a stabilizing action on the vitreous network of the glass, leading to an increase in synthesis yields, but their concentration cannot be directly correlated with the variation of synthesis yield. The sol–gel synthesis technique in the study and the technological routes adopted can be successfully used to obtain calcium phosphate glasses from ternary and quaternary compositional systems, allowing the introduction of the fourth oxide component carrying the dopant metal.

The results obtained, through the values of the synthesis yields higher than those provided by the specialized literature, confirm the versatility of the method and the high level of reproducibility. The overall conclusion of the DLS (Dynamic Light Scattering) analysis of the samples of calcium phosphate glasses synthesized by the sol–gel technique is that all of them are biocompatible, being able to reach micron and even nanometer sizes through further processing (grinding, ultrasonication).

References

1. Varmette, E.A., Nowalk, J.R., Flick, L.M., Hall, M.M.: Abrogation of the inflammatory response in LPS-stimulated RAW 264.7 murine macrophages by Zn- and Cu-doped bioactive sol–gel glasses. J. Biomed. Mater. Res. (2008). https://doi.org/10.1002/jbm.a.32098
2. Cioffi, N., Torsi, L., Ditaranto, N., Tantillo, G., Ghibelli, L., Sabbatini, L., Bleve-Zacheo, T., D'Alessio, M., Zambonin, P.G., Traversa, E.: Copper nanoparticle/polymer composites with antifungal and bacteriostatic properties. Chem. Mater. **17**, 5255–5526 (2005). https://pubs.acs.org/doi/10.1021/cm0505244
3. Joosten, U., Joist, A., Frebel, T., Brandt, B., Diederichs, S., Von Eiff, C.: Evaluation of an in situ setting injectable calcium phosphate as a new carrier material for gentamicin in the treatment of chronic osteomyelitis: studies in vitro and in vivo (2004). https://doi.org/10.1016/j.biomaterials.2003.10.083
4. Liu, W.C., Wong, C.T., Fong, M.K., Cheung, W.S., Kao, R.Y.T., Luk, K.D.K.: Gentamicin-loaded strontium-containing hydroxyapatite bioactive bone cement—an efficient bioactive antibiotic drug delivery system. J. Biomed. Mater. Res. B. Appl. Biomater. (2010). https://doi.org/10.1002/jbm.b.31730

5. Hesaraki, S., Nemati, R.: Cephalexin-loaded injectable macroporous calcium phosphate bone cement. J. Biomed. Mater. Res. B Appl. Biomater. (2009). https://doi.org/10.1002/jbm.b.31222
6. Sasaki, T., Ishibashi, Y., Katano, H., Nagumo, A., Toh, S.: In vitro elution of vancomycin from calcium phosphate cement. J. Arthroplasty (2005). https://doi.org/10.1016/j.arth.2005.03.035
7. Jiang, P.J., Patel, S., Gbureck, U., Caley, R., Grover, L.M.: Comparing the efficacy of three bioceramic matrices for the release of vancomycin hydrochloride. J. Biomed. Mater. Res. B. Appl. Biomater. (2010). https://doi.org/10.1002/jbm.b.31557
8. Sugawara, A., Asaoka, K., Ding, S.J.: Calcium phosphate-based cements: clinical needs and recent progress. J. Mater. Chem. B. (2013). https://doi.org/10.1039/c2tb00061j
9. Russel, A.D., Hugo, W.B.: Antimicrobial activity and action of silver. Prog. Med. Chem. (1994). https://doi.org/ https://doi.org/10.1016/s0079-6468(08)70024-9
10. Marsh, P.D.: Microbiological aspects of the chemical control of plaque and gingivitis. J. Dent. Res. (1992). https://doi.org/10.1177/00220345920710071501
11. Knowles, J.C.: Phosphate based glasses for biomedical applications. J. Mater. Chem. **13**, 2395–2401 (2003)
12. Antoniac, I.: Biologically Responsive Biomaterials for Tissue Engineering. Springer Series in Biomaterials Science and Engineering (2013)
13. Avram, D., Angelescu, N., Ungureanu, D.N., Ionita, I., Gheboianu, A., Bancuta, I.: Study of phosphocalcic glasses SiO_2-CaO-P_2O_5 system with and without silver I. Synthesis of glasses and characterization by WD- XRF and XRD, Revista de Chimie (2017). https://doi.org/10.37358/RC.17.5.5586

Chapter 9
Study of the Calcium Phosphate Glasses Doped with Silver and Copper II. Study of Bioactivity by XRD Analysis

Daniela Avram, Dan Nicolae Ungureanu, Ionica Ionita, Cosmina Cristiana Avram, and Aurora Anca Poinescu

Abstract The bioactivity of the synthesized calcium phosphate glasses, they were immersed in simulated body fluid (SBF) for 3, 7, 14 and 21 days. The structural changes that appear on the surface of the glasses synthesized and subsequently soaked in simulated human fluid were studied by X-ray diffraction. All glasses demonstrate their bioactivity after only three days of immersion in simulant body fluid (SBF), reaching a maximum of bioactivity after 14 days of immersion. The X-ray diffraction (XRD) investigation aimed to identify the crystalline phases associated with the bioactivity process, namely the presence of stoichiometric hydroxyapatite and calcium-deficient hydroxyapatite, carbonates, phosphates, etc. The formation of carbonated hydroxyapatite (HApC) is an advantage for strengthening the bone-implant interfacial bond. The enhanced bioactivity of the doped glasses makes them promising candidates for bone regeneration, facilitating faster healing and improved outcomes.

Keywords Bioglasses · SBF · XRD analysis · Bioactivity · Immersion

D. Avram (✉) · A. A. Poinescu
Department of Food Engineering, Valahia University of Targoviste, Targoviste, Romania
e-mail: daniela.avram@valahia.ro

D. N. Ungureanu · A. A. Poinescu
Department of Material Engineering, Valahia University of Targoviste, Targoviste, Romania

I. Ionita
Department of Sciences and Advanced Technologies, Valahia University of Targoviste, Targoviste, Romania

C. C. Avram
SC Almina Trading SA, Targoviste, Romania

R. Tamas and C. A. Heiman (eds.), *Selected Papers from AToMS 2024*,
Lecture Notes on Multidisciplinary Industrial Engineering,
https://doi.org/10.1007/978-3-032-02452-7_9

9.1 Introduction

The main objective of this study is to analyze how silver and copper ions influence the bioactivity of Ag_2O and Cu_2O doped glasses after soaking in simulated body fluid (SBF), by comparison with ternary glass P1. The analysis of structural changes that appear on the surface of calcium phosphate glasses after immersion in simulated body fluid, different periods of time was highlighted by X-ray diffraction analysis in the scanning range 20–60° 2θ, with a speed of 0.04°/sec.

The XRD analysis was carried out with a diffractometer Rigaku Ultima IV, with the following features: 1αCuk = 1.5405 Å, the source of radiation (X-ray tube with copper anode) has a power of 3 kW and a voltage of 40. The data was collected in scan range 20–60° (2 θ) with a scanning speed of 0.04°/sec. (2 θ).

The specialized literature unanimously confirms that the formation of stoichiometric and non-stoichiometric (carbonated) hydroxyapatite crystalline structures on the surface of biomaterials after immersion in simulated body fluid is associated with their property of being bioactive, interacting beneficially with the host organism [1]. Their bioactivity consists in the ability to form a strong bond at the interface with the bone tissue, with which it comes into contact, through complex physicochemical and biochemical mechanisms that have been described in other studies [2, 3].

The bioactivity of synthesized calcium phosphate glasses was evaluated after immersion in simulated body fluid (SBF) with pH = 7.2–7.25 and thermostated at 37 °C for 3, 7, 14 and 21 days. After each immersion period and before immersion, the samples were investigated by X-ray diffraction. The obtained results are presented in the form of diffractograms on which the newly formed crystalline phases are highlighted.

9.2 Experimental

9.2.1 XRD Analysis of Glass Powders Synthesized by the Solgel Technique

In order to evaluate biocompatibility by in vitro study of the glasses, we synthesized a simulated physiological fluid, according to a recipe of Kokubo and Takadama revised in 2006, called generic simulated body fluid (SBF) [4].

This inorganic, aproteic and acellular solution, with a composition similar to that of biological fluids, proved to be viable, because following the static immersion of the glasses synthesized in it, the de novo synthesis of stoichiometric and carbonated hydroxyapatite was generated on the surface of the materials, a process which, in according to what the specialized literature shows, it is directly correlated with the biofunctionality, more precisely with the bioactivity of the material.

9.3 Results and Discussion

Figure 9.1 shows the diffractogram of the composition samples P1, PC1 and PA1—presented in the first part of the study, which were synthesized by the sol–gel technique and thermally stabilized at 600 °C before immersion in simulated body fluid.

The XRD analysis of the three samples shows the existence of predominantly amorphous compositions characteristic of calcium phosphate glasses, with an incipient crystallization determined by the formation of calcium carbonates and hydroxyapatite crystallites due to a slight devitrification determined by the thermal stabilization treatment.

On the P1 and PA1 diffractograms of glasses, two specific diffraction lines of vaterite (μ-$CaCO_3$) at 2θ are observed: 27.06° and 24.96°, according to the ICDD database—PDF 2 00–033-0286 and two small peaks of hydroxyapatite at 2θ: 43.77° and 49.8°, according to ICDD database—PDF 2 00–009-0432. The XRD analysis of sample P1 also shows a peak specific to calcite (β-$CaCO_3$) at 29.33° 2θ (ICDD–PDF 2 00–005-0586), for sample PA1 a characteristic peak of metallic silver at 2θ 38.07° (ICDD—PDF 2 00–04-0783), and the diffraction spectrum of PC1 glass shows 2 very broad peaks of tenorite (CuO) at 2θ 35.54°, 38.93° and 2 more faint peaks at 2θ 48.77° and 58.31°, according to the ICDD database—PDF 2 00–041-0254.

The presence of calcium carbonates on the surface of the non-immersed glasses may be due to the carbonation of the samples during the stabilization heat treatment at 600 °C because this occurred in the presence of air inside the calcining furnace, not in a controlled atmosphere such as inert gas. The existence of silver at surface of the PA1 sample doped with silver ions is justified by the reduction of a part of Ag_2O in the phosphosilicate matrix of the glass, but this does not exclude the further presence of

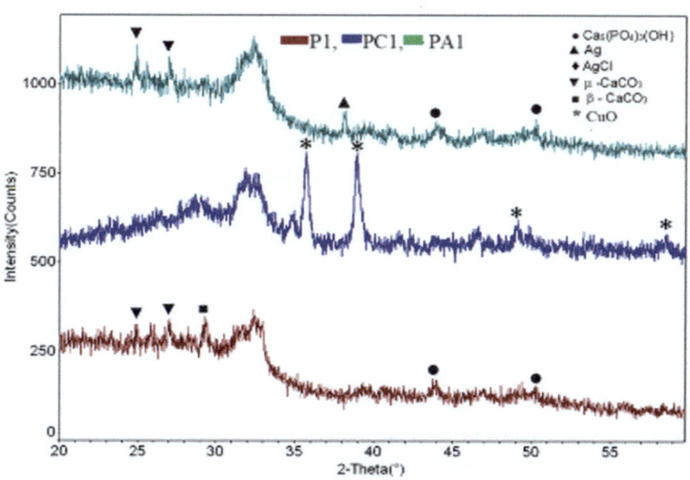

Fig. 9.1 The diffractograms of P1, PC1 and PA1 samples not immersed in SBF

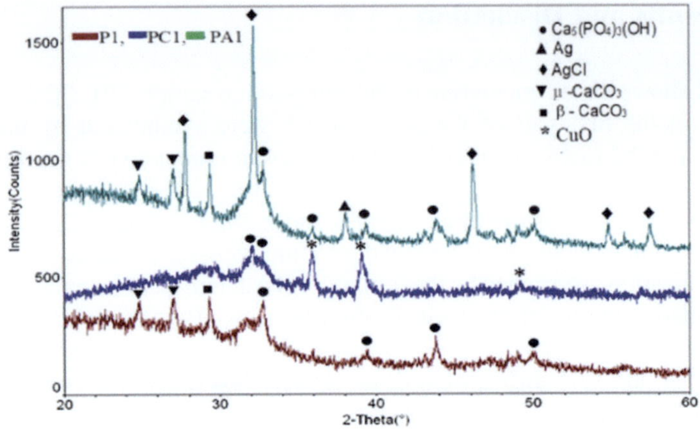

Fig. 9.2 The diffractograms of P1, PC1 and PA1 samples immersed for 3 days in SBF

Ag_2O in the synthesized composition [5]. The glass doped with cuprous ions (PC1) does not present crystalline phases of carbonate or hydroxyapatite type, but it shows 4 characteristic peaks of CuO which was formed as a result of the oxidation of cuprous ions from the copper precursor used in the synthesis (CuCl) and their transformation into cupric ions. This oxidation does not exclude the continued existence in the vitreous matrix of cuprous oxide Cu_2O.

The presence of hydroxyapatite (HAp) on the surface of P1 and PA1 glasses before being immersed in SBF gives them a glass–ceramic type structure, while the PC1 composition preserves its vitreous structure specific to glasses. The metallic silver existing in the vitreous matrix of the PA1 glass gives it, in addition to the other two compositions, a structure specific to glass–ceramic glasses with a composite phase, considering the silver as the reinforcing phase. Figure 9.2 shows the diffractograms of P1, PC1 and PA1 samples after the first 3 days of in vitro testing, by static immersion in SBF, under the conditions specified previously.

Following the XRD analysis of the samples immersed for 3 days, the first surface changes induced by the ion exchange between the glasses and the simulated body fluid are observed, as the diffractograms indicate an increase in the degree of crystallinity evidenced by the formation of new amounts of hydroxyapatite and also by the increase of the size of the already existing HAp crystals (increases the amplitude of the peaks). For P1 sample the changes consist in the appearance of two new peaks of hydroxyapatite at 2θ 32.9° and 39.28° in addition to the existence of the crystalline compounds already present (mentioned previously), which shows that the glass is bioactive. The peaks specific to carbonates also increase in amplitude as a result of immersion in SBF containing bicarbonate ions that can react with calcium ions released from the vitreous matrix of the glass.

The PC1 glass doped also shows 2 new peaks of hydroxyapatite at the diffraction lines 2θ 31.77° and 32.7° which proves that the composition of calcium phosphate glass doped with copper is also bioactive from the first days of immersion and that the doping with ions of copper does not negatively influence the bioactivity of the material. The diffractogram also indicates a change in the Cu^+/Cu^{2+} balance because the CuO peak at the 58.31° 2θ line is no longer visible and the amplitude of the other peaks decreases, possibly due to the reduction of cupric ions to cuprous ions. On the surface of the PA1 sample doped with silver after immersion, major changes in crystallinity appear, consisting in the appearance of three new peaks of hydroxyapatite at 2θ 32.9°, 35.76° and 39.28°, one characteristic of calcite at 2θ 29.31° and four broad AgCl peaks at 2θ: 32.16°, 46.12°, 54.75° and 57.75° according to ICDD database—PDF 2 00–031-1238. The newly formed calcite on the surface of the glass is due to the presence of $NaHCO_3$ in the SBF composition that reacts with calcium ions released from the vitreous matrix, and the formation of large amounts of AgCl crystals is justified by the fact that the reaction between silver ions and Cl-anions is extremely sensitive, it is the basis of argentometry as a chemical method for quantitative dosage of Cl-ions in any system. In addition, this is also explained by the excess of Cl-ions in the composition of SBF (as is also the case with the SBF obtained and used in the in vitro testing), even higher than that in the composition of human plasma (147.8 mmol/L versus 103.0 mmol/L in plasma) [6].

In this respect it can be concluded that after the first days of immersion the calcium phosphate glasses start to show their bioactivity and that silver or copper ion doping of the P1 ternary glass does not compromise the bioactivity of the phosphocalcium material. It can be assumed at this point, considering the number of hydroxyapatite peaks higher in the case of PA1 doped composition, that silver ions could stimulate the reactivity of the material [5]. Figure 9.3 shows the diffractograms of P1, PC1 and PA1 samples after 7 days of immersion in SBF. Compared to the diffractograms in Fig. 9.2, the following changes appear: in the spectrum of P1 sample not doped with metal ions, the increase in the amplitude of the hydroxyapatite peaks in the area 30–35°2θ is observed, with the appearance of a new one at 32.33° 2θ and the disappearance of the vaterite peaks as a result of its solubilization.

It is known that vaterite (β-$CaCO_3$) is much more soluble in water, respectively in simulated body fluid than calcite (β-$CaCO_3$), which can be a beneficial aspect, regarding the global bioactivity of the glass, because the CO_3^{2-} ions released through solubilization can enter the composition of the carbonated hydroxyapatite, which is formed together with the stoichiometric one at the bioglass-SBF interface and which, by embedding collagen proteins in the silicate matrix, accelerates the formation of the tissue-implant interface [7, 8].

In the diffractogram of PC1 copper-doped glass, there are no important changes compared to the diffractogram related to the 3-day immersion period. The CuO peak at 2θ 58.31° identified in the diffractogram of the non-immersed sample is again visible (Fig. 9.1), which indicates a continuous change in the Cu^+/Cu^{2+} balance. Also, the already existing HAp peaks increase in amplitude, proof that its formation continues even after the first 3 days of immersion, under the existing immersion conditions. In the case of PA1 glass doped with silver, it is found the constant maintenance of the quantities of carbonates, AgCl and metallic Ag, but the number of hydroxyapatite (HAp) peaks on the diffractogram is reduced; the peaks of HAp from the positions 35.76° and 39.28° 2θ are no longer visible, a phenomenon that can

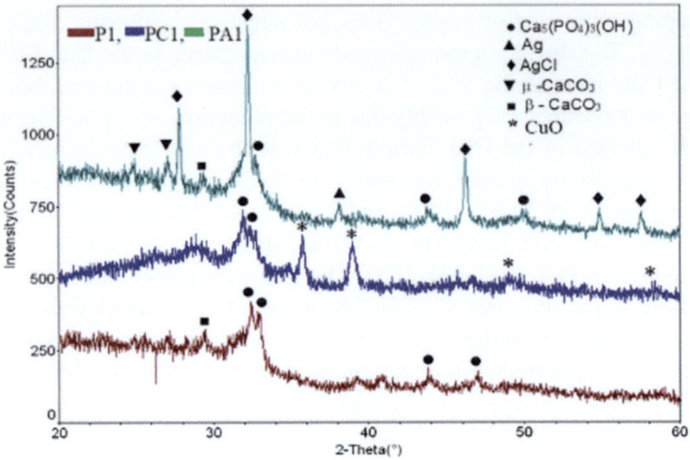

Fig. 9.3 The diffractograms of P1, PC1 and PA1 samples immersed for 7 days in SBF

be explained by the fact that the crystalline AgCl precipitate clogs the pores of the hydrated silica network on the surface of the glass and thus can prevent the formation of hydroxyapatite or, most probably, by blurring the visibility of the HAp peaks as a result of the large amplitude of the AgCl peaks [5, 9].

The diffractograms of the P1, PC1 and PA1 glasses immersed for 14 days in SBF shown in Fig. 9.4 indicates a behavior similar to that of the samples immersed for 7 days, without essential changes in terms of the composition of the crystalline phases, but with important changes in the degree of crystallinity and implicitly of bioactivity.

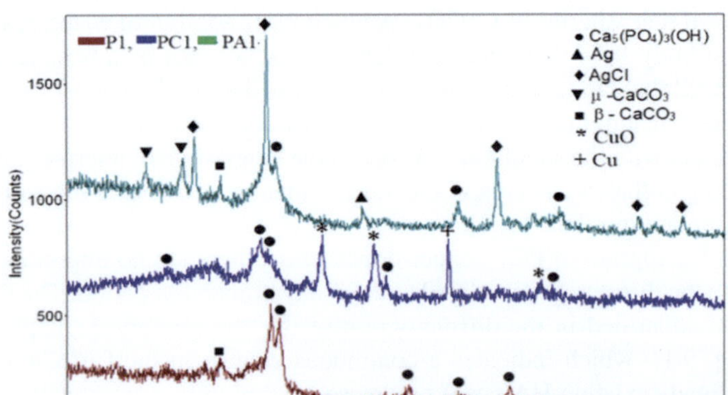

Fig. 9.4 The diffractograms of P1, PC1 and PA1 samples immersed for 14 days in SBF

An increase in the intensity of the diffraction maxima is noted for all crystalline phases, except for carbonates and metallic silver, which denotes the favoring of crystal development with increasing immersion time. The increase in the degree of crystallinity is also due to the increase in the structuring degree of the phases formed post-immersion [9].The diffractogram of the P1 ternary glass indicates the formation of new amounts of hydroxyapatite because the already existing peaks increase in amplitude and a new peak appears at the 2θ 46.96° diffraction line.

In the case of the PC1 composition, the changes are even more obvious because 3 new peaks of hydroxyapatite appear in the positions 2θ 25.9°, 39.4° and 49.44°, but also one of metallic copper at 2θ 43.24° (according to the PDF database 2 00–004-0836), which suggests very intense $Cu/Cu^{+}/Cu^{2+}$ redox phenomena, probably depending on the compositional changes of the simulated body fluid during the duration of immersion and the release rate of cuprous ions from the vitreous matrix. The tenorite peak at 2θ 58.31° is no longer evident. After 21 days of immersing P1, PC1 and PA1 samples in simulated body fluid (XRD analysis from Fig. 9.5), it is found the constant maintenance of the number of peaks of the crystalline phases already formed (calcite, HAp, Ag and AgCl); however, the 32.9° 2θ peak is no longer evident in the diffractogram of P1 sample and the 2 vaterite peaks reappear at the 2θ 24.8° and 26.94° lines, with a much larger amplitude than after immersion for 7 and 14 days.

The association of these processes after long periods of immersion is due to the formation of carbonated hydroxyapatite, a compound that cannot be identified by XRD analysis. This phenomenon can be explained by the fact that due to the static process of immersion calcium and phosphate ions are depleted from the SBF composition or by the fact that the stoichiometric HAp gradually transforms into

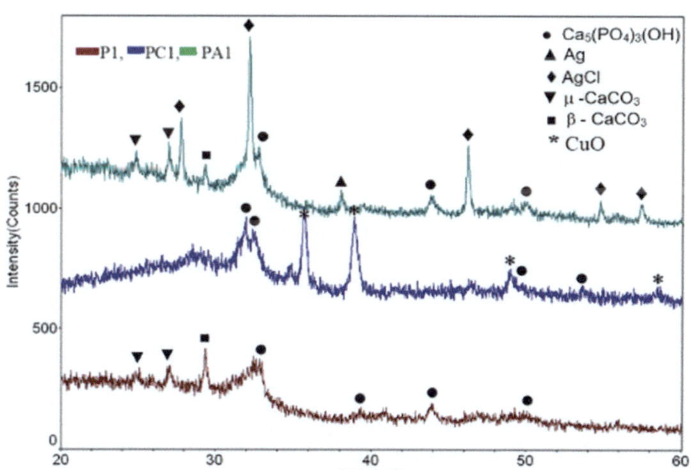

Fig. 9.5 The diffractograms of P1, PC1 and PA1 samples immersed for 21 days in SBF

carbonated hydroxyapatite (HApC) (according to the FTIR analysis presented in other articles), with diffraction line values close to those of vaterite.

The failure of deplete the SBF in calcium and phosphate ions is not actually encountered in real mode in vivo in biological systems except in cases of severe hypocalcemia or major demineralization due to osteoporosis, situations usually incompatible with the surgical act of bone prosthesis [2].

The PC1 quaternary glass has a somewhat similar behavior because also in its case the diffractogram indicates a reduction in the number of apatite peaks (the peak at 2θ 25.9° is no longer visible); the increase in the amplitude of the tenorite peaks and the disappearance of the peak associated with metallic copper, formed after 14 days of immersion, indicate a new change in the balance of copper ions, so a continuous activity at the interface glass—simulated human fluid. The formation of carbonated hydroxyapatite (HApC) is an advantage for strengthening the bone-implant interfacial bond because carbonated hydroxyapatite, non-stoichiometric, is the mineral phase closest to the composition of bone tissue [9, 10].

The XRD analysis provides crucial information about the crystalline structure of the calcium phosphate glasses, revealing its composition and phase. The presence and intensity of specific peaks in the XRD pattern can indicate the degree of bioactivity, providing insight into the glass's potential for bone bonding.

9.4 Conclusions

Calcium phosphate glasses doped with silver and copper offer significant potential in the field of regenerative medicine. Their enhanced bioactivity, biocompatibility, and antimicrobial properties pave the way for innovative solutions in bone regeneration and infection prevention. Future research will focus on optimizing the synthesis, characterization, and application of these advanced materials for clinical applications.

The XRD investigation aimed to identify the crystalline phases associated with the bioactivity process, namely the presence of stoichiometric hydroxyapatite and calcium deficient hydroxyapatite, carbonates, phosphates, etc. The simultaneous presence of several crystalline phases in the structure of some calcium phosphate glasses synthesized for this research study (before immersion in SBF), leads to their inclusion in the category of ceramic glasses and glasses with a composite phase. After the first 3 days of immersion, the calcium phosphate glasses begin to show their bioactivity: the doping of glasses with silver ions does not compromise the bioactivity of phosphocalcium materials, but on the contrary, silver ions could stimulate the reactivity of the material. The incorporation of silver and copper introduces antimicrobial properties, making the glass suitable for implants in areas prone to infection. After 14 days of immersion, the diffractograms of all glasses synthesized by the sol–gel technique indicate a significant increase in the number of HAp peaks on the surface of all compositions, associated with an increase in the degree of bioactivity (maximum level).

The glass doped with cuprous ions PC1 does not present crystalline phases of the carbonate or hydroxyapatite type, before immersion in SBF, but it exhibits peaks characteristic of tenorite (CuO), which was formed because of the oxidation of Cu^+ ions from the copper precursor used in the synthesis (CuCl) and their transformation into cupric ions Cu^{2+}.

Advantage of the XRD investigations lead to the general conclusion that all the glasses synthesized and analyzed by the sol–gel technique, ternary and those doped with silver or copper, are bioactive and generate new hydroxyapatite formations post-immersion in simulating human fluid on their surface as early as the first 3 days and reach a maximum of bioactivity after 14 days of immersion. This study demonstrated that the bioactive glasses synthesized in this research are biocompatible. Doping with silver or copper ions of P1 ternary glass does not cancel or negatively influence the bioactivity and biocompatibility of the synthesized material, on the contrary, it gives them (PA1, PC1) and antibacterial properties [1, 11].

References

1. Avram, D., Angelescu, N., Ungureanu, D.N., Ionita, I., Gheboianu, A., Bancuta, I.: Study of phosphocalcic glasses SiO_2–CaO–P_2O_5 system with and without silver I. Synthesis of glasses and characterization by WD- XRF and XRD. Revista de Chimie (2017). https://doi.org/10.37358/RC.17.5.5586
2. Avram, D., Angelescu, N., Ungureanu, D.N., Ionita, I., Bancuta, I., Gheboianu, A., Lungulescu, E.M.: Study of phosphocalcic glasses SiO_2–CaO–P_2O_5 system with and without silver II. The bioactivity analysis by FTIR, SEM methods and microbiological study of silver-doped glasses. Revista de Chimie (2017). https://doi.org/10.37358/RC.17.6.5639
3. Avram, D., Angelescu, N., Ungureanu, D.N., Ionita, I., Popescu, E.C.: The bioactivity assessment of silver-doped phosphocalcic glasses. Adv. Top. Optoelectron., Microelectron., Nanotechnologies, SPIE **10977** (2018). https://doi.org/10.1117/12.2323596
4. Kokubo, T., Takadama, H.: How useful is SBF in predicting in vivo bone bioactivity? Biomaterials **27** (2006). https://doi.org/10.1016/j.biomaterials.2006.01.017
5. Perez Gonzalez, J.A.: Síntesis de biovidrios por la técnica sol-gel con incorporación de metales y estudio de sus propiedades antibacteriales. PhD thesis, Universidad de Chile, Facultad de Ciencias Físicas y Matemáticas, Departamento de Ingeniería Química y Biotecnología (2012)
6. Vulpoi, A.: Studii privind biomateriale compozite. PhD thesis summary, Univ. Babes-Bolyai, Cluj, Romania (2011)
7. Ungureanu, D.N., Angelescu, N., Avram, D.: Synthesis, characterization and in vitro bioactivity of SiO_2–CaO–P_2O_5 sol—gel glasses highlighted by XRD technique. The Scientific Bulletin of Valahia University Materials and Mechanics, no.6 (2011)
8. Ungureanu, D.N., Angelescu, N., Catangiu, A., Stoian, E.V., Rizescu, C.Z., Avram, D.: Synthesis and characterization of bioactive sol–Gel glass: a preliminary study. Adv. Mater. Res. (2012). https://doi.org/10.4028/www.scientific.net/AMR.341-342.21
9. Hench, L.L., Wilson, J.: Bioceramics. Mater. Res. Soc. Bull. **16** (1991)
10. Dumitrascu, N.: Biomateriale şi biocompatibilitate. Universitatii Alexandru Ioan Cuza, Iasi (2007)
11. Avram, D., Ungureanu, D.N., Angelescu, N., Gheboianu, A., Bancuta, I., Bratu, M.G.: Study of bioactivity and antimicrobial activity in case of glasses from SiO_2–CaO–P_2O_5 ternary system, p. 10. The Scientific Bulletin of Valahia University, Materials and Mechanics (2015)

12. Antoniac, I.: Biologically responsive biomaterials for tissue engineering. In: Series: Springer Series in Biomaterials Science and Engineering, vol. 1. Springer (2013)
13. Hench, L.L., West, J.K.: Biological applications of bioactive glasses. Life Chem. Rep. (1998). https://doi.org/10.4236/njgc.2013.32011

Chapter 10
The Influence of the Chrome Layer Thickness on the Physical–Mechanical Properties of Steels

Aurora Anca Poinescu, Ionica Ionita, Ana-Maria Hossu, Daniela Avram, and Simona Mihai

Abstract The work reveals the importance of the chromium layer deposited on steel to improve corrosion resistance. This corrosion resistance of the chrome material depends, among other things, on the thickness of the chrome layer. The chrome layer, in turn, is influenced by the parameters of the chrome plating bath (the intensity of the currents used, the advance of the material through the chrome plating baths, the temperature of the bath, the concentration of the solutions, etc. Undesirable problems such as matt deposition, milky white chromium deposition, insufficient thickness and hardness are the most common problems faced in the electroplating industry (Bayramoglu et al. in J. Mater. Process. Technol. 203:277–286, 2008). The material subjected to adhesion, microscopy, and hardness is a 38MnVS6 alloy steel used to make the rod of hydraulic cylinder. The influence of deposited layer thickness was also investigated on the corrosion resistance in salt fog.

Keywords Chromium layer · Salt fog

A. A. Poinescu (✉)
Department of Material Engineering, Valahia University of Targoviste, Targoviste, Romania
e-mail: aurora.poinescu@valahia.ro

I. Ionita · A.-M. Hossu
Department of Sciences and Advanced Technologies, Valahia University of Targoviste, Targoviste, Romania

D. Avram
Department of Food Engineering, Valahia University of Targoviste, Targoviste, Romania

S. Mihai
Institute of Multidisciplinary Research for Science and Technology, Valahia University of Targoviste, Targoviste, Romania

R. Tamas and C. A. Heiman (eds.), *Selected Papers from AToMS 2024*,
Lecture Notes on Multidisciplinary Industrial Engineering,
https://doi.org/10.1007/978-3-032-02452-7_10

10.1 Introduction

The electrolytic deposition of chromium is a complex process and despite the large number of attempts carried out, there is still no complete explanation of some phenomena, so the cathodic effect of foreign anions (catalysts) is not fully elucidated with all the attempts at explanation [1]. The safe part is the formation in solution of some compounds that contain trivalent chromium together with chromate and sulfate and which play a decisive role on the cathode film [2–4]. The purpose of electrolytic chromium deposition is to reduce the wear of moving parts. By using new materials, surface coatings, improving lubricants and lubrication systems, energy losses due to friction and wear could be reduced by 40% in the long term [5].

Chromium electrodeposition coating is formed by chemical or electrochemical treatment of metals or metal coatings in solutions containing hexavalent chromium (Cr^{6+}) and other components. The process results in the formation of an amorphous protective layer composed of the substrate, complex chromium compounds and other substances. The properties of this type of electrolytic deposition are hardness, good corrosion and abrasion resistance, friction resistance and finally, the decorative appearance.

Chrome deposits have good heat resistance and start to change color at 480-500°C. The hardness of chrome-coated surfaces exceeds the hardness of the hardest steels. Compared to other electroplating technology, chrome plating has a series of distinct features in the sense that the electrolyte is a little sensitive to impurities and obtaining satisfactory deposits is conditioned by the presence of some ions in the electrolyte. The baths of the chroming plant contain chromic acid which is a donor of chromium ions, and an acid radical called a catalyst which improves the cathodic deposition of chromium. The anodes of the chroming plant are made of lead, tin, and antimony alloy. The current density must be between 10 and 100 A/dm^2. The purpose of electrolytic chromium deposition is to reduce the wear of moving parts.

The crystallographic structure of the chromium deposits is cubic with centered faces. In the cross-section attacked with the Murakami reagent, the structure is characterized by the presence of cracks in the form of closed channels whose number and size are variable. The physical and mechanical properties of galvanic deposits are influenced by factors such as the composition and nature of the electrolyte, the presence of superficially active substances, temperature, current density, etc. These properties are appreciated by determining microhardness, brittleness, internal stresses, and corrosion resistance. Internal tensions are compression or expansion forces that appear in the structure of the electrolytically deposited metal layer. The appearance of internal stresses is due to the changes in the parameters of the crystalline network, the change in the distance between the crystals of the deposit. The size of the internal voltage depends on factors such as the nature of the support material, the nature of the deposited material, the thickness of the deposited layer, the temperature, the density of the electrolyte, the acidity of the solution. The manifestation of internal stress can lead to a decrease in anti-corrosion properties because of cracking of the coatings or exfoliation from the base material [6].

The thickness of the chrome layer will directly depend on the working parameters: electrolyte concentration, catalysts, working temperature, current efficiency, working advance. For deposits that exceed 0.03 mm, a second chrome plating is used to reach the thickness of the deposit specified by the customer in the order. Requirements for electrolytic coatings with a high chromium layer $\geq 40\ \mu m$ are from the manufacturers of hydraulic cylinders used in hostile, corrosive environments, for example trucks equipped with the rear closing of the lift-type body that have an intensive degree of use and a potential environment corrosive (garbage trucks). Another field of use of this type of hydraulic cylinder is that of marine drilling. Marine drilling platforms are complex installations that operate in a hostile environment [7].

According to the authors Protsenko and Danilov [8], the electroreduction of Cr (III) complex ions in aqueous solutions proceeds gradually, through the formation of intermediate Cr (II) compounds, and the rate of chromium deposition is determined by the current density of the Cr (II) ion electroreduction [9–11].

10.2 Material and Methods

The paper highlights the importance of electrochemical coatings with chromium, a process that improves the corrosion resistance and mechanical properties of steels. This corrosion resistance of the chrome material depends, among other things, on the thickness of the chrome layer. The chromium layer is in turn influenced by the parameters of the chroming bath (intensity of the currents used, advance of the material through the chroming baths, temperature of the bath, concentrations of the solutions), etc. The analyzed material is a 38MnVS6 brand steel used for the manufacture of the hydraulic cylinder rod. The bar with a diameter of 50 mm was chromed and polished.

The sanding operation carried out with abrasive belts and brushes in combined media (dry, wet) aims to improve surface quality (elimination of non-conformities) and improve surface roughness within the limits required by the finishing operation (0.2 μm). Roughness has an important influence on seal performance, low friction, and seal life. A rough or very smooth surface led to negative results. Surface microcracks allow very small volumes of fluid to pass under the seal during each operating cycle.

This lubricating film reduces friction and extends seal life. At the opposite pole, the very thin surface reduces the amount of lubricants, and at the seal-stem interface there is more friction and therefore more wear. The wear rate, i.e. the number of operating cycles, is a key factor in dynamic applications and depends on the quality of the rod and the seal.

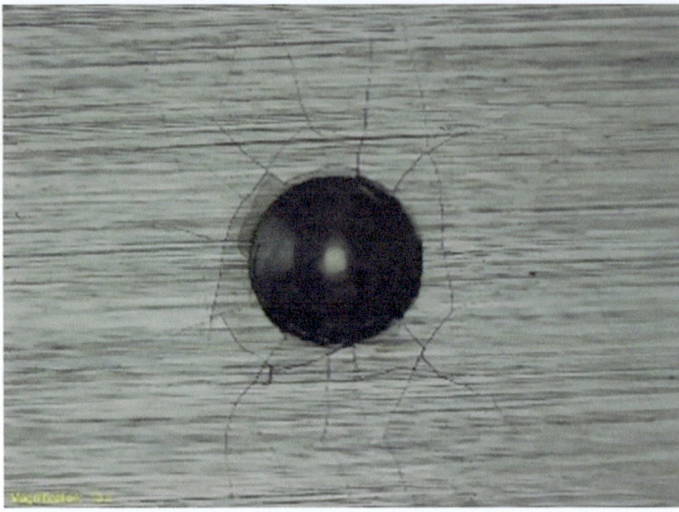

Fig. 10.1 HRC Indentation micrograph, X100

10.3 Results and Discussion

10.3.1 Chrome Coating Adhesion Test

The chromium layer adhesion test is performed on the chromium surface by Rockwell indentation test (HRC), applying a force of 150 N. The method of assessing the adhesion/strength of the chromium layer is done by comparing the micrographs of the sample analyzed with the standards of the VDI 3198 method, shown in the following Fig. 10.1.

Following the adhesion/strength test of the chromium layer we can conclude that the sample has satisfactory characteristics and can be classified as HF4 according to the method used.

10.3.2 Hardness Test of the Chrome Layer

The hardness test of the chrome layer was performed according to SR EN 6507–1 Vickers method using the HV0.1 microhardness test with an applied force of 0.890 N. The AFFRI DM2 device was used to measure hardness by the Vickers method. According to the standard, they made three indentations, their average being 987 HV0.1, and in Fig. 10.2 you can see the marks of Vickers hardness for the 3 determinations made.

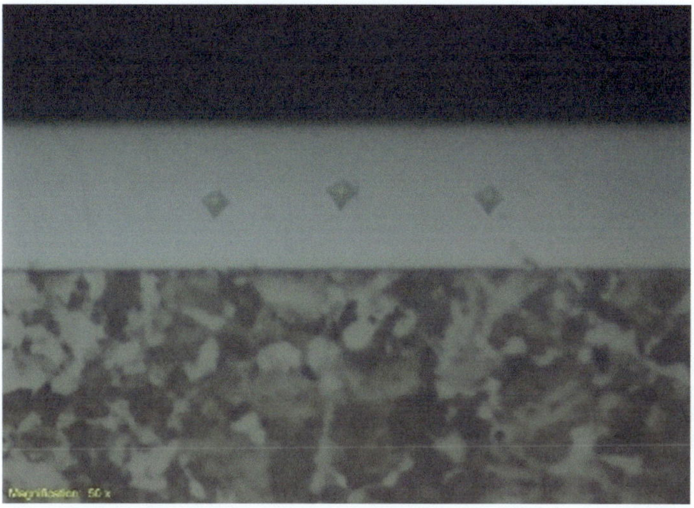

Fig. 10.2 Vickers indentation HV0.1, X500

10.3.3 Microscopic Analysis

Through microscopic analysis, the metallographic structure of the 38MnVS6 steel and the thickness of the chrome layer are highlighted. After the preliminary preparation of the samples (grinding, polishing and chemical attack), they were analyzed with a Olympus GX53 metallographic optical microscope. Figure 10.3 shows both the basic structure of the steel, a pearlitic ferrite structure, and a 57 μm chromium layer.

Microcracks in the chromium layer can be used to link the external environment to the base material. This is a trigger for corrosion under the chromium layer with negative consequences for the whole cylinder assembly, which in a short time will become nonfunctional. To avoid this type of defect, passivation it is necessary. Passivation consists of depositing a thin film, adherent to the surface and penetrating through microcracks, thus sealing the metal material. The passive film offers high stability in corrosive environments, uniformity and low porosity. The determining factors of the passivation process are the current density and the electrochemical potential of the passivation solution.

The attack of the sample with the Murakami reagent highlights a multilayer chromium layer deposition, the first layer measures approximately 3 μm, the second layer 31.19 μm and the third 27.08 μm, as well as the microcracks in the cross section (Fig. 10.4).

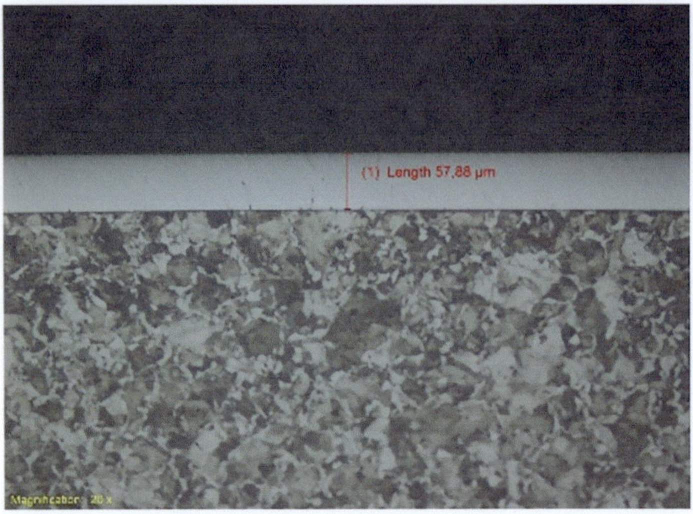

Fig. 10.3 38MnVS6 steel microstructure, X200, Cr layer 57.88 μm

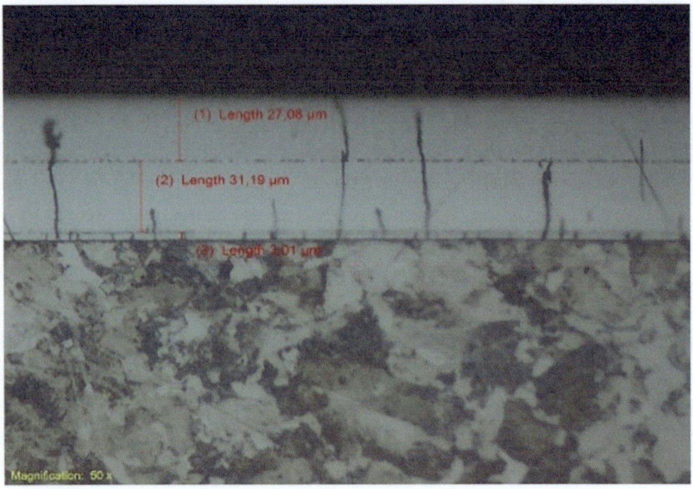

Fig. 10.4 38MnVS6 steel microstructure, X500, Murakami attack chrome layer

10.3.4 The Salt Spray Test

The samples that will be subjected to the corrosion test were visually checked to identify any defects on the surface. The sample to be tested is:—degreased,—the roughness is measured—the deposited chrome layer—isolate with adhesive tape the upper and lower part of the sample, the ends resulting from cutting—the sample is labeled

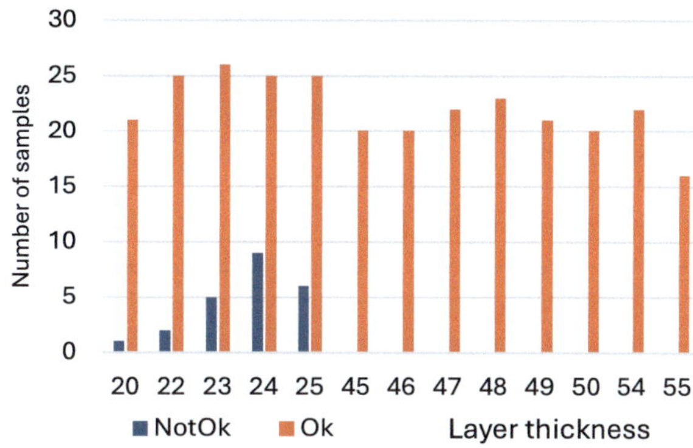

Fig. 10.5 The salt fog resistance test of chrome samples

with the identification data of the material (type, diameter, batch, rating, number of test hours, date of introduction into the test chamber)—the prepared sample is introduced into the salt fog chamber, it is sealed and the accelerated corrosion test device is turned on.

During the performance of the test, every 24 h the values of the parameters certifying the degree of corrosion of the test chamber during operation are determined: the concentration of the solution; the pH of the solution; saline room temperature; the pluviometry constant.

At hourly intervals of 24 h, the device is turned off, the samples are taken out of the saline fog chamber, washed under a jet of water to remove rust leaks from the surface of the sample and are evaluated from the point of view of corrosion resistance. Following the evaluation of the samples, a rating is given (qualifying from 1 to 10). The sample prepared for evaluation is compared with the corrosion diagrams that have an established rating (according to ISO 10289/2001). Thus, the appearance of the sample is compared with that of the corrosion diagrams, the rating corresponding to a corrosion diagram is also given to the tested sample.

The tested samples have a chromium layer between 18 and 55 μm and received a score, (rating) depending on the behavior of each one during the corrosion resistance test compared to the product rating.

Following the salt fog resistance test, we can conclude that out of the total samples analyzed, samples with a chromium layer between 20 and 25 μm had a higher number of non-compliant samples than those with a chromium layer between 46 and 55 μm, as can be seen from the graph in Fig. 10.5.

10.4 Conclusions

The structure of the base material, the thickness of the deposited layer, the uniformity of the electrochemical deposition, the lack of defects that may appear between the base material and the chrome layer were highlighted through microscopic analysis.

The adhesion test of the chrome layer performed according to EN ISO 2819:2017-point 4.15 Rockwell test, VDI 3198 method, places the sample in the HF4 adhesion/resistance class, satisfactory adhesion if we refer to the thickness of the deposited layer. The measurement of the hardness of the chromium layer was carried out by the Vickers method using the HV0.1 microhardness test with a force of 0.980 N, the result obtained is 987 HV0.1. It is known that the hardness of the chromium layer depends on the size of the grains [12]. The structure of the 38MnVS6 alloy steel is a pearlite-ferritic structure.

The attack of the sample with the Murakami reagent highlights a deposition of the multilayer chromium layer. The multilayer deposition type improves corrosion resistance and increases the service life.

Corrosion in salt fog for the samples with a chromium layer between 20–25 μm reveal the fact that not all the samples are compliant instead the samples with chromium layer between 46 and 55 μm, the salt fog resistance is much improved. The samples that were passivated after chromium plating performed much better in the salt fog test.

References

1. Aubert, A., Gillet, R.: Hard chrome coatings deposited by physical vapour deposition. Thin Solid Films **108**, 165–172 (1983). https://doi.org/10.1179/030716983803291343
2. Irudaya Raj, L.M., Sathishkumar, J., Kumaragurubaran, B., Gopal, P.: Analysis of hard chromium coating defects and its prevention methods. Int. J. Eng. Adv. Technol. (IJEAT) **2**(5). ISSN: 2249 – 8958 (2013). www.ijeat.org
3. Kuppusami, A.D., Raghunathan, V.S.: A new surface treatment by pulsed plasma nitriding for chromium plated austenitic stainless steel. ISIJ Int. **42**(12), 1457–1460 (2002). https://doi.org/10.2355/isijinternational.42.1457
4. Wu, B., Fu, Y., Xu, J., et al.: Chromium nitride films on stainless steel as bipolar plate for proton exchange membrane fuel cell. J. Power Sources **194**, 976–980 (2009). https://doi.org/10.1016/j.jpowsour.2009.06.029
5. Protsenko, V.S., Danilov, F.I., Gordiienko, V.O., et al.: Improving hardness and tribological characteristics of nanocrystalline Cr–C films obtained from Cr (III) plating bath using pulsed electrodeposition. Int. J. Refract. Metals Hard Mater. **31**, 281–283 (2012). https://doi.org/10.1016/j.ijrmhm.2011.10.006
6. Podgornik, B., Massler, O., et al.: Crack density and tribological performance of hard-chrome coatings. Tribol. Int. **121**, 333–340 (2018). https://doi.org/10.1016/j.triboint.2018.01.055
7. Zhang, Q., JiangangWang, W.S., et al.: Failure analysis of chromium plating layer on the surface of the piston rod of the hydraulic jack. Coatings **12**, 774 (2022). https://doi.org/10.3390/coatings12060774
8. Protsenko, V., Danilov, F.: Kinetics and mechanism of chromium electrodeposition from formate and oxalate solutions of Cr (III) compounds. Electrochim. Acta **54**, 5666–5672 (2009). https://doi.org/10.1016/j.electacta.2009.04.072
9. Zhao, H., Liu, W., Li, Q., Zhang, B. et al.: Mechanism of chromium electrodeposition from Cr (III) baths on nickel and chromium electrode surfaces. Int. J. Electrochem. Sci. **15**, 8979–8989 (2020). https://doi.org/10.20964/2020.09.23
10. Eugénio, S., Rangel, C.M., et al.: Electrodeposition of black chromium spectrally selective coatings from a Cr (III)–ionic liquid solution. Thin Solid Films **519**, 1845–1850 (2011). https://doi.org/10.1016/j.tsf.2010.10.029

11. Molodkina, E.B., Ehrenburg, M.R., et al.: Electrodeposition of chromium on single-crystal electrodes from solutions of Cr (II) and Cr (III) salts in ionic liquids. J. Electroanal. Chem. **860**, 113892 (2020). https://doi.org/10.1016/j.jelechem.2020.113892
12. Choi, Y., Kim, M., Kwon, S.C.: Characterization of chrome layer formed by pulse plating. Surf. Coat. Technol. **169–170**, 81–84 (2003). https://doi.org/10.1016/S0257-8972(03)00159-2
13. Bayramoglu, M., Onat, B., Geren, N.: Statistical optimization of process parameters to obtain maximum thickness and brightness in chromium plating. J. Mater. Process. Technol. **203**, 277–286 (2008). https://doi.org/10.1016/j.jmatprotec.2007.09.063

Chapter 11
Numerical Simulation of Ballistic Impact of a 1/1 Type Aramid Fabric Panel at Meso-Level

Larisa Titire and Cristian Munteniță

Abstract The aim of this paper is to analyze the behavior of a 24-layer aramid fabric, 1/1 type aramid panel under projectile impact and to examine how the variation of the aramid yarn yield strength can influence the results of ballistic impact. The numerical simulations allow for a detailed analysis of these influences, thus providing important data for the development of more effective protective materials and the optimization of flexible panels for impact resistance. To gain a clear insight into how variations in aramid yarn yield strength influence the behavior of the textile panel during impact with a 9 mm FMJ projectile, we performed simulations for 11 different values of aramid yarn yield strength.

Keywords Numerical simulation · Yield strength · Aramid · Impact ballistic · 9 mm FMJ

11.1 Introduction

Ballistic protection has evolved significantly in recent years, particularly in the materials and design of soft ballistic inserts [1]. The soft armor, made of flexible ballistic materials, is designed to protect against handguns and is light in weight, making it suitable for prolonged daily wear [2]. These inserts consist of multiple layers, which influence the energy absorption capacity [3].

Ballistic inserts are made of para-aramids with a woven structure in which the weft and warp threads intersect at right angles, creating a criss-cross pattern [4]. These yarns are made of braided fibers that have a high ratio between their length and cross-section. Due to its impact resistance, lightweight character and high energy absorption capabilities, this material is widely used in ballistic applications. These

L. Titire (✉) · C. Munteniță
Faculty of Mechanical Engineering, University "Dunarea de Jos", Galați, Romania
e-mail: larisa.chiper@ugal.ro

R. Tamas and C. A. Heiman (eds.), *Selected Papers from AToMS 2024*,
Lecture Notes on Multidisciplinary Industrial Engineering,
https://doi.org/10.1007/978-3-032-02452-7_11

characteristics make it particularly valuable for the manufacture of bulletproof armor [5].

In their study, Chu et al. [6] examined the mechanical characteristics of the yarns. It was determined that the density of the yarn does not have a substantial impact on the ballistic performance of the cloth. In contrast, it was noted that a higher value of longitudinal Young's modulus results in a more rapid deceleration of the projectile.

In their study, Rao et al. [7] demonstrated that material properties, boundary conditions and friction play an important role in the ballistic performance of the fabric. The results show that friction between yarns improves ballistic performance more than friction between projectile and fabric, especially for stiffer and stronger yarns. When impact velocity higher than the ballistic limit, friction between projectile and fabric is more beneficial than friction between yarns. Fabrics made of high stiffness yarns decelerate the projectile faster, while high strength yarns delay the onset of damage. The combination of stiffness and maximum strength gives the best ballistic performance.

The study and analysis of the behavior of composite materials against ballistic impact is a crucial area in the development of protection and security technologies. In this context, the evaluation of the performance of materials, such as aramid yarn with different yield strengths, plays a fundamental role in determining their ability to withstand extreme stresses and to protect life and property against ballistic threats [8].

This paper focuses on the analysis of results obtained in numerical impact simulations of composite structures using various yield strengths of aramid yarn. Each investigated case provides a detailed insight into how the material responds to impact, highlighting stress distributions, deformations and fracture behaviors in successive layers.

11.2 The Model

The ballistic panel is made of 1/1 layered fabric. The thickness of the fabric is approximately 0.6 mm and the size of the panel is 48 mm × 48 mm. The panel is embedded on all ends and is subject to impact by a 9 mm FMJ (Full Metal Jacket) projectile at a velocity of 420 m/s, according to the NIJ Standard [9], classifying it as protection level II. The cross-section of the yarn is lenticular in shape [8]. Yarns that are in direct contact with the projectile are called main yarns and yarns that are not in direct contact with the projectile are named secondary yarns.

The contact conditions take into account the friction phenomenon only for the contact between yarn and projectile, using a coefficient of friction (COF) of 0.25 [8]. The friction between the yarns is negligible in order to reduce the computation time and to allow the simulation of the panel without symmetry planes. This approach was chosen because the fabric is modelled at meso-level and the large number of bodies (yarns) in contact requires an increased computational capacity. In these simulations, the Lagrangian formula is used to reproduce the behaviour of both the projectile and the woven panel in 1/1 configuration.

The material properties involved in the impact are selected from the literature [10–12]. Each case has the same initial conditions, impact conditions, the same material properties for the projectile. In the case of the yarn properties, the only difference is the value for the yarn yield strength, the other values remain constant for all cases. The scale of the model is meso-level because it involves the yarns as a single homogeneous isotropic material. Yarn density is 1440 kg/m3, Young's Modulus is 90000 MPa, Poisson's ration 0.35, Tangent Modulus is 1000 MPa. The projectile core density is 11340 kg/m3, Young Modulus 16,000 MPa, Poisson's ratio is 0.44. Density of the projectile's jacket is 8300 kg/m3, Young's Modulus 1.17×105 MPa, Poisson's Ratio is 0.34, Yield Strength is 70 MPa and Tangent Modulus 1150 MPa. The failure criterion for the materials involved in the impact is EPS (Equivalent Plastic Strain).

The 11 simulated models are noted with case 1 for 2000 MPa yarn yield strength, case 2 for 2100 MPa thread yield strength, up to case 11 with 3000 MPa thread yield strength. The aramid yarn yield strength increases from case to case by 100 MPa.

Based on the equivalent stresses and images obtained during the simulation, the impact behavior of the two variants is analyzed. In the first case analyzed, the yarn yield strength analyzed is 2000 MPa (case 1) and the last case has a yarn yield strength of 3000 MPa (case 11).

11.3 Results and Discussion

Understanding how the variation in the yield strength of aramid fibers influences the mode of deformation within a ballistic protection panel is important for optimizing its design for personal protection systems. When aramid fibers with a higher yield strength are incorporated into a ballistic protection panel, the panel tends to show more elastic deformation when subjected to impact. Aramid fibers with a lower yield strength are more prone to plastic deformation when subjected to stress. Plastic deformation involves permanent changes in the shape of the material, usually beyond the elastic limit. In applications where personal protection is the primary concern, higher yield strength aramid fiber panels may be preferred (they can provide effective protection by absorbing energy while reducing the risk of injury to the wearer due to deformation).

In case 1, at the end of the numerical simulation, a Von Mises stress of 546 MPa is recorded, localized on the tip of the projectile (Fig. 11.1). The impact causes complete breakage of all layers by the projectile, which is also deformed by the impact. The final layers show a more significant damage, caused by both the deformation of the projectile and the intense compression of the material during the impact. On the tip of the projectile, the presence of broken fragments of the yarns can be observed, emphasizing the intensity of the impact and its effects on the structure of the composite materials.

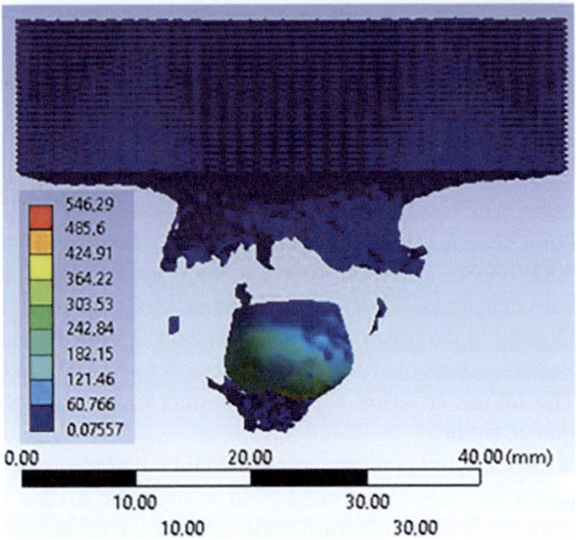

Fig. 11.1 Case 1, yield strength = 2000 MPa

In case 2, at the end of the numerical simulation, a maximum Von Mises stress value of 594 MPa is recorded, also located on the tip of the projectile (Fig. 11.2). In this scenario, the projectile undergoes a trajectory change during impact, which influences the stress distribution and material behavior. Similar to case 1, broken fragments of the yarn are observed on the bullet tip. The change in trajectory affects the stress distribution and structural strength of the composite material.

Case 3, with the aramid yarn yield strength of 2200 MPa, shows a maximum Von Mises stress of 618 MPa in the numerical simulation (Fig. 11.3). During impact, the projectile changes its trajectory, which contributes to the intense stress distribution

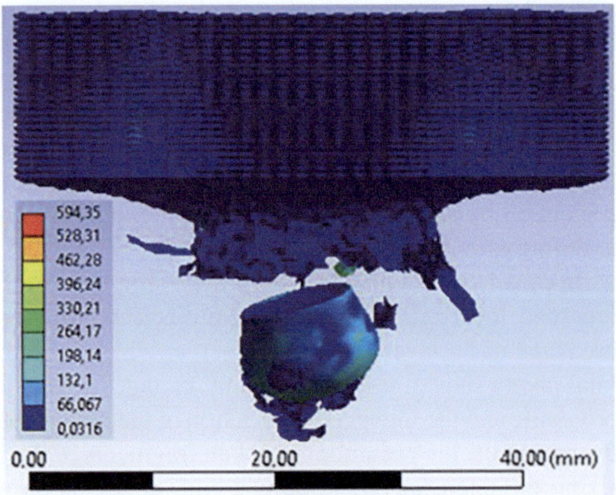

Fig. 11.2 Case 2, yield strength = 2100 MPa

in the material. All layers are broken in this process, indicating severe deformation and maximum absorption of impact energy throughout the structure.

Case 4, with the aramid yarn yield strength of 2300 MPa, at the end of the numerical simulation records a Von Mises stress value of 653 Mpa (Fig. 11.4). During impact, the projectile undergoes a change in trajectory, which contributes to the intense stress distribution in the material. On the tip of the projectile, broken fragments of the yarns in the impact zone are observed.

In case 5, with a yarn yield strength of 2400 MPa, the numerical simulation shows the damage of all layers (Fig. 11.5). The Von Mises stress value recorded is 1936 MPa, and broken fragments of the yarn are present on the projectile tip. The bullet manages to break all the layers, but does not completely exit the fabric, due to the

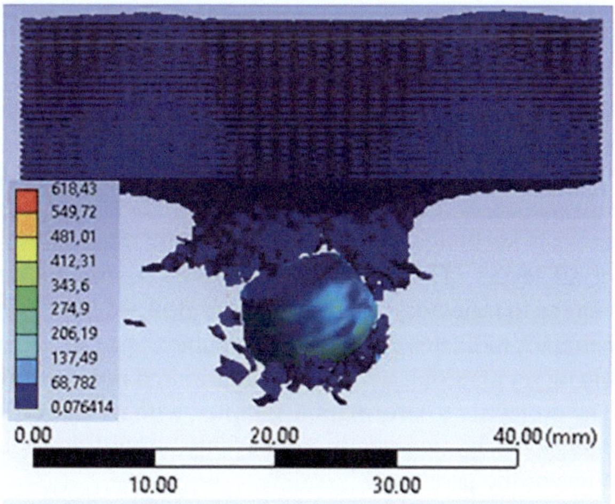

Fig. 11.3 Case 3, yield strength = 2200 MPa

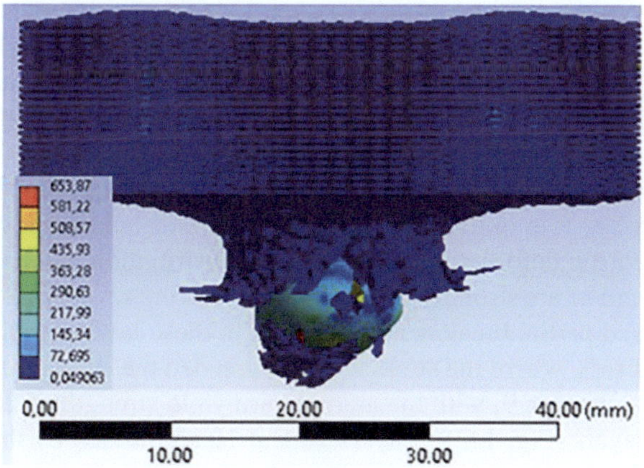

Fig. 11.4 Case 4, yield strength = 2300 MPa

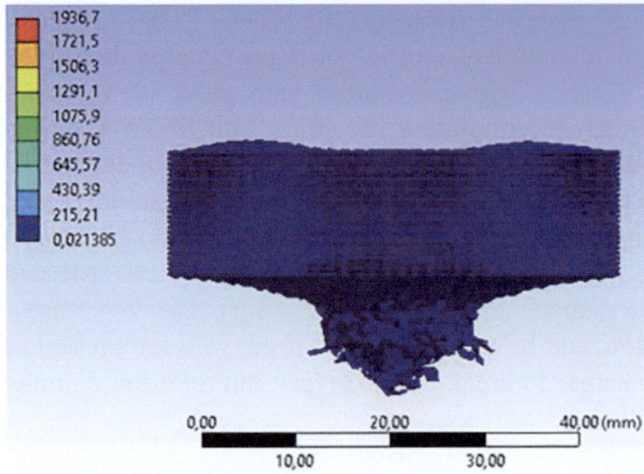

Fig. 11.5 Case 5, yield strength = 2400 MPa

increased strength of the yarns. This may require extending the numerical simulation time to fully capture the material behavior.

In Case 6, with the aramid yarn yield strength of 2500 MPa, a higher von Mises stress is observed at the end of the numerical simulation (Fig. 11.6). This increased stress is attributed to the continued fragmentation of the yarn even after the damage of all layers. The projectile undergoes significant deformation and experiences a change in trajectory upon impact, resulting in asymmetric damage of the layers. This behavior indicates a heightened complexity in the interaction between the bullet and the target material, with effects distributed non-uniformly across the impact surface. The projectile's deformation, coupled with its altered trajectory, suggests that kinetic energy dissipation occurred unevenly, amplifying localized damage within certain areas of the layers.

In case 7, with the aramid yield strength of 2600 MPa, the projectile is stopped (Fig. 11.7). However, the layers under the projectile are compressed and show broken yarns in the impact zone. Layer 23 and layer 24 are not penetrated.

In case 8, with the aramid yarn yield strength of 2700 MPa, the projectile is stopped, but the number of destroyed layers is smaller (Fig. 11.8). This case shows 3 unpenetrated layers. Layer 19, although broken in the impact zone, contributes to the projectile rebound. Layers 20 and 21 are heavily compressed, with completely and partially broken yarns. Layer 19, although broken in the impact zone, played a key role in the projectile rebound. This behaviour could be attributed to both the elastic deformation and the stress distribution mechanism in the layers. Layers 20 and 21 are strongly compressed, indicating severe plastic deformation. The complete and partial breakage of the yarns in these layers emphasizes the areas of maximum stress, where the stresses have exceeded the strength limits of the material.

In case 9, with the aramid yarn yield strength of 2800 MPa, there are 19 broken layers (Fig. 11.9). Layers 18 and 19 are broken by the projectile, but the projectile fails to advance very far. These two layers are destroyed by compression. The number of broken layers reflects the intensity of the impact and the energy transferred by the

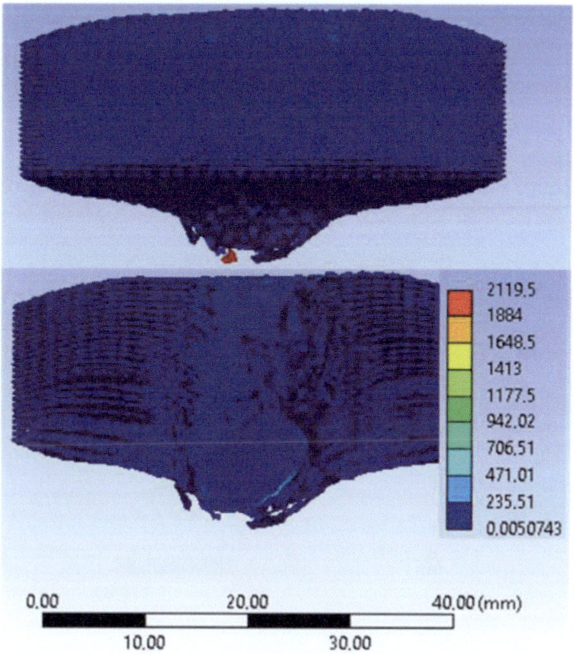

Fig. 11.6 Case 6, yield strength = 2500 MPa

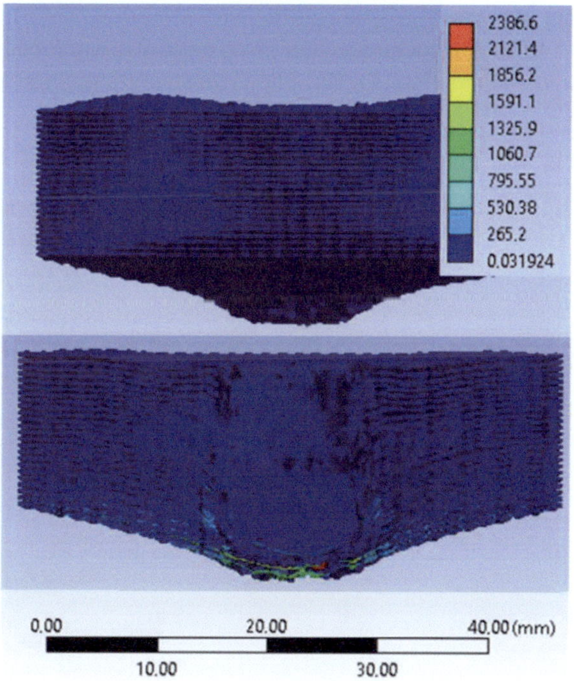

Fig. 11.7 Case 7, yield strength = 2600 MPa

Fig. 11.8 Case 8, yield strength = 2700 MPa

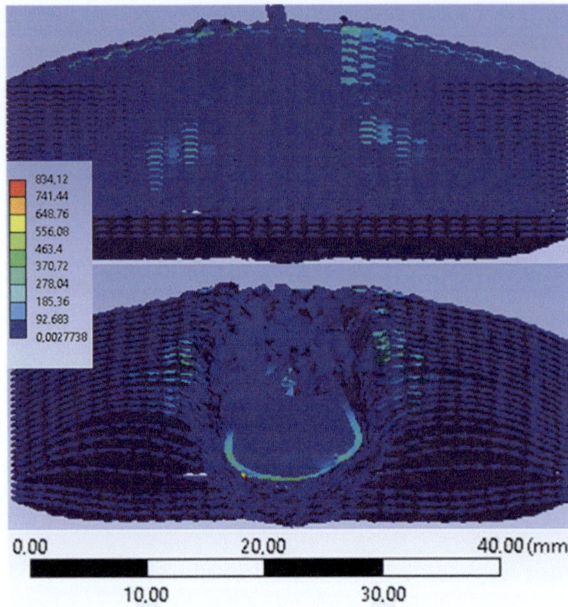

projectile to the target material. In spite of this damage, the material managed to stop the projectile, suggesting efficient absorption of the impact energy.

In case 10, with the aramid yarn yield strength of 2900 MPa, there are 6 unpenetrated layers (Fig. 11.10). The yield strength of 2900 MPa represents an exceptional

Fig. 11.9 Case 9, yield strength = 2800 MPa

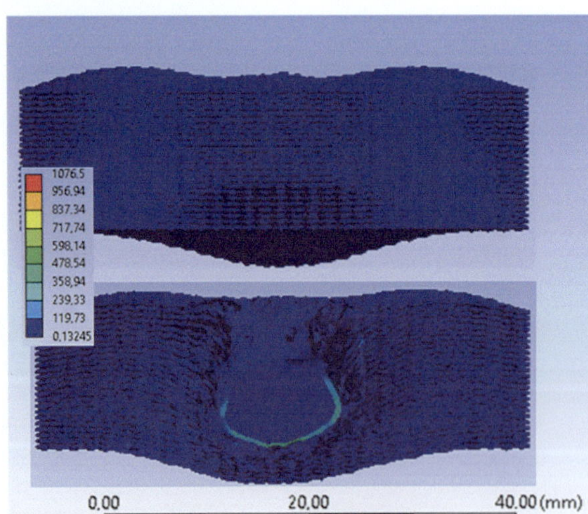

strength of the material, which contributes to an increased ability to stop the projectile and limit the breakage of the layers. The presence of 6 unpenetrated layers indicates superior projectile stopping performance, demonstrating that the material can disperse and absorb impact energy very effectively.

Case 11, with a yarn yield strength of 3000 MPa, records 7 unpenetrated projectile layers (Fig. 11.11).

Fig. 11.10 Case 10, yield strength = 2900 MPa

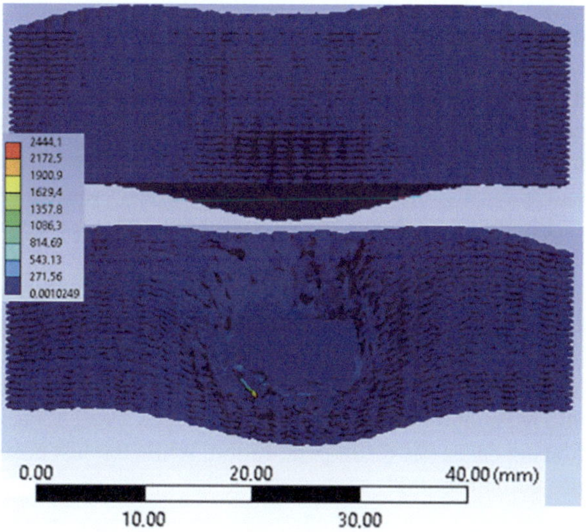

Fig. 11.11 Case 11, yield strength = 3000 MPa

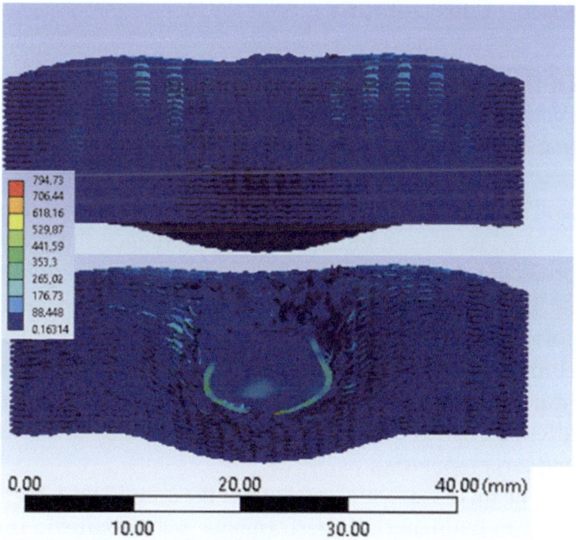

The energy lost by the projectile as it passes through the fabric is equal to the energy absorbed by the target material in the area targeted by the projectile. During impact, the fabric impact position, and over time the wave begins to propagate to neighboring and adjacent points along the axis of the yarn, while the deformation at the point of initial stress increases.

A shear wave, compression wave and in-plane deformation propagation will all be formed simultaneously in the thickness direction by the movement of the region below the projectile tip. The stress in the warp and weft yarns forming the cone during impact in the main tensioned region will arise from the compressive stress, which will cause a conical deformation below the projectile tip. If the stresses on the yarns are greater than the breaking stresses, the yarns will fail under the stresses created by compressing, stretching, or shearing the yarns. Secondary yarns will resist movement of the main yarns due to friction between yarns at the point of intersection and between adjacent yarns which increases tensile forces.

Figure 11.12 shows graphs for the equivalent stress of the eleven analyzed cases. At the beginning of the numerical simulation, each case records an increasing equivalent stress, which is influenced by the value of the yield strength of the yarn. The equivalent stress remains constant during the first half of the numerical simulation, after which it starts to oscillate. The first four cases record values between 550 and 650 MPa at the end of the simulation. The sixth and eleventh cases record values around 800 MPa, and the remaining cases exceed 1900 MPa, reaching up to 2687 MPa. The first six cases show full penetration, while the last cases show partial penetration and projectile rebound.

11.4 Conclusions

Of the 11 cases studied, 5 showed partial penetration and 6 total penetrations (all layers damaged). Case 7 has 2 unpenetrated layers after impact, case 8 has 3 unpenetrated layers, case 9 has 5 unpenetrated starts, case 10 has 6 unpenetrated layers and case 11 has 7 unpenetrated layers.

The deformation of the projectile upon impact indicates that its material has reached its plastic limit. This may influence the trajectory and effectiveness of the projectile, leading to less effective penetration and higher kinetic energy dispersion.

The observation of broken yarn fragments on the tip of the projectile suggests a complex interaction between the projectile and the target material, where shear and compressive forces contribute to the destruction of the material. The change in the projectile trajectory during impact is an important phenomenon that can significantly affect ballistic behavior. This is influenced by the stress distribution in the projectile and target material.

The increased yield strength of the aramid yarn highlights the significant improvement in the material's performance under ballistic impact conditions. The unpenetrated layers suggest that the material can withstand the stresses without failing

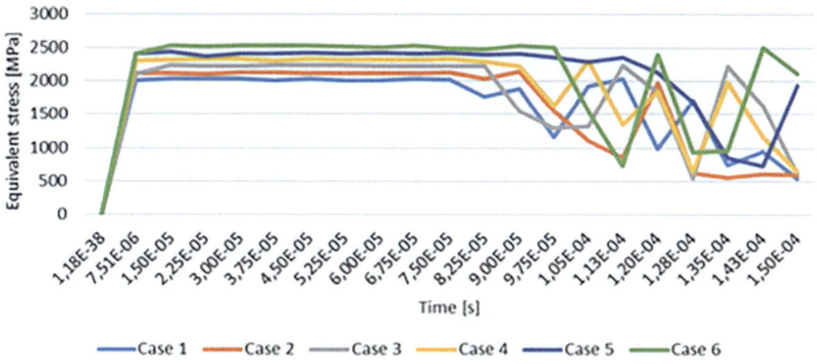

(a) Equivalent stress, case 1-6

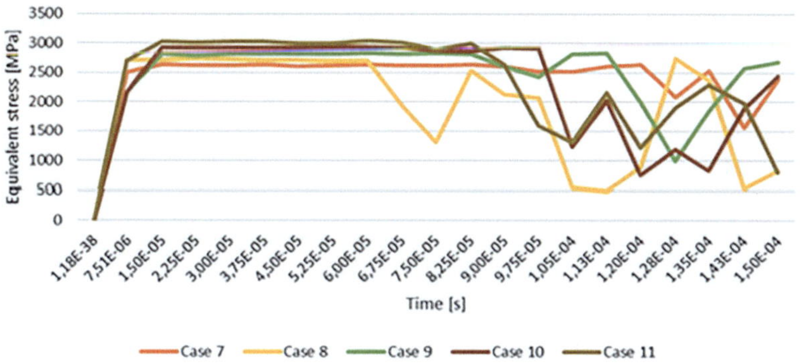

(b) Equivalent stress, case 7-11

Fig. 11.12 The equivalent stress distribution

completely. The use of aramid yarn with higher yield strengths may provide superior protection and significantly reduce the damage under impact conditions.

References

1. Abtew, M.A., Boussu, F., Bruniaux, P., Loghin, C., Cristian, I.: Ballistic impact mechanisms–a review on textiles and fi-bre-reinforced composites impact responses. Compos. Struct. **223**, 110966 (2019)
2. Zochowski, P., Bajkowski, M., Grygoruk, R., Magier, M., Burian, W., Pyka, D., Bocian, M., Jamroziak, K.: Finite element modeling of ballistic inserts containing aramid fabrics under projectile impact conditions–comparison of methods. Compos. Struct. **294**, 115752 (2022)
3. Kośla, K., Kubiak, P., Łandwijt, M., Urbaniak, W., Kucharska-Jastrząbek, A.: Fragment-resistant property optimization within ballistic inserts obtained on the basis of para-aramid materials. Materials **15**, 2314 (2022)
4. Feito, N., Loya, J.A., Muñoz-Sánchez, A., Das, R.: Numerical modelling of ballistic impact response at low velocity in aramid fabrics. Materials **12**, 2087 (2019)

5. Kędzierski, P., Gieleta, R., Morka, A., Niezgoda, T., Surma, Z.: Experimental study of hybrid soft ballistic structures. Compos. Struct. **153**, 204–211 (2016)
6. Chu, T.L., Ha-Minh, C., Imad, A.: A numerical investigation of the influence of yarn mechanical and physical properties on the ballistic impact behavior of a Kevlar KM2® woven fabric. Compos. Part B Eng. **95**, 144–154 (2016)
7. Rao, M.P., Duan, Y., Keefe, M., Powers, B.M., Bogetti, T.A.: Modeling the effects of yarn material properties and friction on the ballistic impact of a plain-weave fabric. Compos. Struct. **89**(4), 556–566 (2009)
8. Chiper Titire, L.: Numerical and experimental study for ballistic protection systems with aramid fibre fabrics. PhD thesis, "Dunarea de Jos" University, Galați, România (2023)
9. NIJ Standard–0101.06: Ballistic resistance of personal body armor. U.S. Department of Justice Office of Justice Programs National Institute of Justice: Washington, DC, USA (2008)
10. Børvik, T., Dey, S., Clausen, A.H.: Perforation resistance of five different high-strength steel plates subjected to small-arms projectiles. Int. J. Impact Eng **36**, 948–964 (2009)
11. Peroni, L., Scapin, M., Fichera, C., Manes, A., Giglio, M.: Mechanical properties at high strain-rate of lead core and brass jacket of aNATO 7.62 mm ball bullet in numerical simulations of ballistic impacts. Proc. DYMAT **26**, 01060 (2012)
12. Tan, V.B.C., Zeng, X.S., Shim, V.P.W.: Characterization and constitutive modeling of aramid fibers at high strain rates. Int. J. Impact Eng **35**, 1303–1313 (2008)

Chapter 12
Generation of Photonic "Molecules" Under the Action of Ultraviolet Light in the DNA/RNA Dimerization Process

Nicolae A. Enaki, Diana Podoleanu, Ion Munteanu, Igor Costisen, Elena Starodub, and Tatiana Paslari

Abstract A semi-classical model (classical by the field and quantum by describing biomolecules) is proposed, which establishes the dependence of the DNA/RNA dimerization rate on ultraviolet C (UVC) radiation intensity. In particular, a nonlinear model based on cooperative processes similar to the Raman-like in quantum optics is developed. The main result of the theory demonstrates that the dimerization process in DNA/RNA strongly depends on the UVC light intensity, thus proving a possible quantum microscopic mechanism of UVC light interactions with DNA. A notion of a photonic molecule consisting of the normal state and the dimer state under the action of the bimodal Raman field is given. This type of molecule exists only in the presence of the excited field, and after it is turned off, the biomolecule remains part of the time in the entangled state after which DNA dimerization is probabilistically established. A nonlinear dependence on the intensity of the applied radiation was established, which was confirmed in a series of experiments on the inactivation of pathogens (or yeast colonies) reproduction under the action of ultraviolet radiation. The main objective of this study was to establish a link between the intensity of UV-C radiation and the efficiency of decontamination using metamaterials with an increased contact surface with contaminated fluid by pathogens, that cause nosocomial infections, namely Escherichia coli, Pseudomonas aeruginosa, Candida albicans, and Candida glabrata. These microorganisms were put through different forms of packing, both thick fibers repacked with thin fibers and millimeter balls repacked with sub-millimeter quartz balls. The results demonstrate that bacterial pathogens respond differently to exposure to UV-C radiation, depending on the contact surface created between the elements of the quasi-periodic metamaterials.

N. A. Enaki (✉) · D. Podoleanu · I. Munteanu · I. Costisen · E. Starodub · T. Paslari
Quantum Optics and Kinetic Processes Lab Institute of Applied Physics, Moldova State University, Chisinau, MD 2028, Republic of Moldova
e-mail: enakinicolae@yahoo.com

R. Tamas and C. A. Heiman (eds.), *Selected Papers from AToMS 2024*,
Lecture Notes on Multidisciplinary Industrial Engineering,
https://doi.org/10.1007/978-3-032-02452-7_12

12.1 Introduction

The possibility of creating localized states linked between cellular partners was an idea proposed by Kasha and is actually the basis of the formation of some supracellular structures in which the excited state is relatively stable in time and the electronic subsystem does not recombine in time. Such supracellular patterns could arise upon the cooperative absorption of photons in the multiple Raman scattering process. This model assumes that between the normal state and the excited state of the bipolar molecules it is possible to form a localized state of the electrons of both molecules so that the system has a relatively long lifetime. If we talk about a system of such bipolar molecules upon excitation, they can enter into a bipolar bond that could lead to the formation of relatively large molecules. A model of multiple excitation of molecular states and the formation of molecular clusters in induced Raman scattering is proposed in this chapter. The model is reduced to the multiple steps of absorption and emission of photons in the scattering process.

The model could be used in photosynthesize using natural and artificial photoreceptors, mostly of microbial origin, to be genetically introduced into the cells of interest. For the multiple Raman effect to have an increased efficiency, optical cavities are needed. In biological systems, such optical cavities are observed. For example, in photosynthesis, both the chloroplast membrane and its internal elements are used as a resonator. The chloroplast is a structure in the cells of plants and green algae that is the site of photosynthesis, the process by which light energy is converted into chemical energy, resulting in the production of oxygen and energy-rich organic compounds. With the introduction of photosensitive opsin/rhodopsin it was observed that cells that are naturally insensitive to light can be made photosensitive and addressable by illumination and can be precisely controlled in time and space. If we look at some proteins as radiation accumulation resonators, then possibilities of selective Raman directing when introducing such proteins into the cell can take place at the subcellular level. In other words, the cellular subsystems are activated by applying control elements such as promoters, enhancers and specific targeting sequences to the DNA/proteins encoding the photoreceptor used. This powerful approach enables precise characterization and manipulation of cellular functions and has motivated the development of advanced optical methods for patterned photostimulation.

As follows from the literature [1, 2] a nucleosome is the basic structural unit of DNA packaging in eukaryotes and may serves not only support for DNA cain but the possible electromagnetic contact through the evanescent zone of the elements of this repached structure in nucleus with dimension about 5- 10 μm. The structure of a nucleosome consists of a segment of DNA wound around eight histone proteins [1, 2] and resembles thread wrapped around a spool. The nucleosome is the fundamental subunit of chromatin. Each nucleosome is composed of a little less than two turns of DNA wrapped around a set of eight proteins called histones, which are known as a histone octamer. Each histone octamer is composed of two copies each of the histone proteins H2A, H2B, H3, and H4 (see Ref. [3–5]). Due to the fact that the refractive index of histone elements is higer than the water one it is posible to brelize the standung or guide waves in such a system like in the traditional optical emements.

In ADN modification, the histones are highly basic proteins abundant in lysine and arginine residues that are found in eukaryotic cell nuclei and play an imortant role in the localization of vibration waves as this take place in microspherical resonators (see Fig. 12.1) B. They act as spools around which DNA winds to create structural units called nucleosomes [3, 4]. Nucleosomes in turn are wrapped into 30- nanometer fibers that form tightly packed chromatin. Histones prevent DNA from becoming tangled and protect it from DNA damage. In addition, histones play important roles in gene regulation and DNA replication. Without histones, unwound DNA in chromosomes would be very long. For example, each human cell has about 1.8 m of DNA if completely stretched out; however, when wound about histones, this length is reduced to about 90 μm (0.09 mm) of 30 nm diameter chromatin fibers [5]. In other words, the subsystems of the cell nucleus in eukaryotes are activated by applying control elements, such as promoters, enhancers and specific targeting sequences, to the DNA/proteins that encode the photoreceptor used. This powerful approach enables precise characterization and manipulation of cellular functions and has motivated the development of advanced optical methods for patterned photostimulation. A huge catalogue of histone modifications have been described, but a functional understanding of mostis still lacking. Collectively, it is thought that histone modifications may underlie a histone code, whereby combinations of histone modifications have specific meanings. However, most functional data concerns individual prominent histone modifications that are biochemically amenable to detailed study.

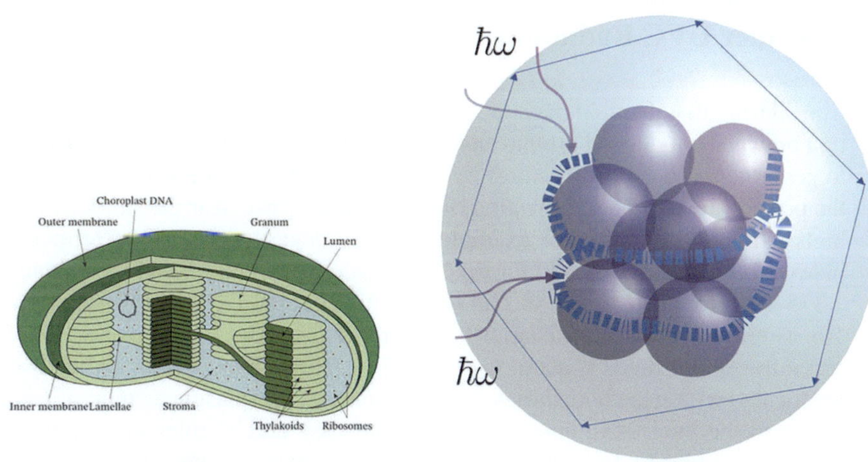

(a) (b)

Fig. 12.1 **a** Observation of disc-shaped optical microresonators (photoreceptors) in chloroplasts in the process of storing and using radiation in photosynthesis. **b** Observation of spherical Hystone elements in the DNA packing [1–5]

In ADN modification, the histones are highly basic proteins abundant in lysine and arginine residues that are found in eukaryotic cell nuclei and play an imortant role in the localization of vibration waves as this take place in microspherical resonators (see Fig. 12.1) B. They act as spools around which DNA winds to create structural units called nucleosomes [3, 4]. Nucleosomes in turn are wrapped into 30-nm fibers that form tightly packed chromatin. Histones prevent DNA from becoming tangled and protect it from DNA damage. In addition, histones play important roles in gene regulation and DNA replication. Without histones, unwound DNA in chromosomes would be very long. For example, each human cell has about 1.8 m of DNA if completely stretched out; however, when wound about histones, this length is reduced to about 90 μm (0.09 mm) of 30 nm diameter chromatin fibers [5]. In other words, the subsystems of the cell nucleus in eukaryotes are activated by applying control elements, such as promoters, enhancers and specific targeting sequences, to the DNA/proteins that encode the photoreceptor used. This powerful approach enables precise characterization and manipulation of cellular functions and has motivated the development of advanced optical methods for patterned photostimulation. A huge catalogue of histone modifications have been described, but a functional understanding of most is still lacking. Collectively, it is thought that histone modifications may underlie a histone code, whereby combinations of histone modifications have specific meanings. However, most functional data concerns individual prominent histone modifications that are biochemically amenable to detailed study.

12.2 Modification of Biomolecular Structural Changes Under UVC Radiation

A. Considering that the field may generate such a bistable state relative the excitation between unbounded ground and bonded supramolecular states, we propose to generate these supra-structures by Raman excitation process. Let us discuss the possibilities of the stabilization of the standing wave near the surfaces with higher refraction index. For example in many eukaryotic cells the refractive indexes is about 1.365 1.390 for thitoplasme/cytosol; 1.355 1.365 for nucleus; 1.375 1.385 for nucleolus; 1.400 1.420 for mithocondria; 1.600 for lissosome [6–10]. These indexes is relative large than the refractive index of water $n_0 = 1.3325$ and intracellular fluids is about $n_{0f} = 1.35$ [6, 7]. Both cytoplasm and cytosol are located within the boundaries of the cell membrane (plasma membrane) and surround the nucleus in eukaryotic cells and play crucial roles in supporting cellular activities by providing a medium for chemical reactions to occur. This optical material may be used as a good internal reflective material, reflecting the light waves into the nucleus/cell if the incident angle is less than Snell's quantified angle relative to the separation surface of the two optical media. This effect may localize the standing wave near the surface of such a quasi spherical cavity similar to gallery waves, so that it may induce the coherent Raman excitation of the proposed above bound molecular structures induced by multi-step scattering Hamiltonian (see semiclassical approach in Refs. [11–13] and quantum aspects in Refs. [14–16]. Denoting by the excitation, $\hat{R}_i^+$, and lowering, $\hat{R}_i^-$, operators in right direction of the chain, and by the similar operators the dipole positive defined direction of the chain of the dipoles constructed by tile to cap connections, the chain detected in opposite sense, $\hat{L}_i^+$ and $\hat{L}_i^-$, we may define the Hamiltonian for the creation of proteins molecules under the action of bimodal cavity field (see Refs. [14–16]. In bad cavity limits for classical bimodal field in interaction with two-level system which represents the bonded and unbounded molecular dipole components described by the master equation,

$$\frac{\partial}{\partial}\hat{W}(t) = \frac{i}{h}\left[\hat{W}(t), \hat{H}_I\right] + \hat{\mathcal{L}}\hat{W}(t) \tag{12.1}$$

where the Hamiltonian part of master equation is,

$$\begin{aligned}\hat{H}_I = {} & h\Delta_r \sum_i \hat{R}_i^z + h\Delta_l \sum_i \hat{L}_i^z - h\sum_{i,j} \chi_{ij}^{(r)} \hat{R}_i^+ \hat{R}_j^- \\ & - h\sum_{i,j} \chi_{ij}^{(l)} \hat{L}_i^+ \hat{L}_j^- - h\sum_{i,j} \chi_{ij}^{(lr)} \left(\hat{L}_i^+ \hat{R}_j^- + \hat{R}_j^+ \hat{L}_j^-\right) \\ & + h\Omega_r \sum_i \left(\hat{R}_i^+ + \hat{R}_i^-\right) + h\Omega_l \sum_i \left(\hat{L}_i^+ + L_i^-\right),\end{aligned} \tag{12.2}$$

and the part of master Eq. (12.2), $\hat{\mathcal{L}}\hat{W}(t)$ describes the losses from cavity. Here the interaction between the neighbor protein subsystem takes place, $\sum_l \chi_{ij}^{(r)} \hat{R}_i^+ \hat{R}_j^- = \chi_{ii-1}^{(r)} \hat{R}_i^+ \hat{R}_{i-1}^- + \chi_{ii+1}^{(r)} \hat{R}_i^+ \hat{R}_{i+1}^-$. The similar sum is represented and for the left dipole circuit, $\sum_j \chi_{ij}^{(l)} \hat{L}_i^+ \hat{L}_j^- = \chi_{ii-1}^{(l)} \hat{L}_i^+ \hat{L}_{i-1}^- + \chi_{ii+1}^{(l)} \hat{L}_i^+ \hat{L}_{i+1}^-$; $\Omega_r = \kappa \sum_{j=1}^{k} g_k \widehat{\wedge_{k}^{-}}_{r} \exp[i\omega t]$, and

$\Omega_l = \kappa \sum_{j=1}^{k} g_k \widehat{\wedge_{k}^{-}}_{l} \exp[i\omega t]$ are the Rabi frequencies of interaction of the bimodal cavity field in the R- and L filaments in interaction with each other by the expression; $\sum_j = i \pm 1 \chi_{ij}^{(lr)} \hat{L}_i \hat{R}_j^-$; $\Delta_r = \omega_r - \omega$; $\Delta_l = \omega_l - \omega$, where ω_r, and ω_l are the frequencies of excitation in the right and left protein filaments represented in Fig. 12.2; In the both ringing we may do the diagonalisation of the spectrum so that for left and right circuits contain the new spectrum. Considering that the field acts only on the right rings we represent these rings as a periodically arranged between the tertian component of not formed yet left rings represented in Fig. 12.2,

$$h\Delta_r \sum_i \hat{R}_i^z - h\sum_{i,j} \chi_{ij}^{(r)} \hat{R}_i^+ \hat{R}_j^- = -\frac{h\Delta_r |GG|}{2}$$

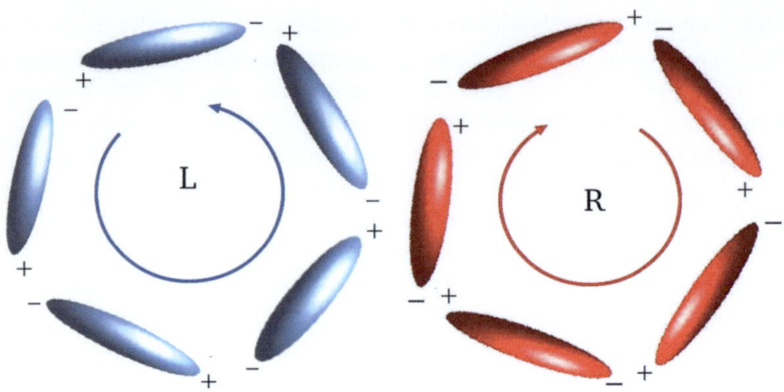

Fig. 12.2 The Planar multi-mers represented above may have the left or right polarizations as this is represented above

$$+\sum_{\alpha=1}^{n}\left[\frac{h\Delta_r}{2}+(E_\alpha-I)\right]|E_\alpha E_\alpha|;$$

Here for penta-mer we have the energies, $E_{1,2}-I_1=w(1+\sqrt{5})/2$; $E_{1,2}-I_1=w(1$ $\sqrt{5})/2$ $(E_{1,2}\,I_1=w(1+5)/2$, and $E_5-I_1=-2w$. We are interested in the generation of supramolecular relative the ground state E_5-I_1 the left rings stimulated by right ones. For this we introduce the condition that the Ranan resonance relative excite state of bounded molecular to satisfy the condition; $\frac{h\Delta_r}{2}(E_n I)\frac{h\Delta_r}{2}$; for tri-mer, quadri-mer, and penta-mer this condition is, $\Delta_r=2w/\hbar$. Introducing the new re-normalized molecular operators we obtain the master equation which takes into consideration the interaction of two filament/rings structures,

$$\frac{\partial}{\partial t}\hat{W}(t)=\left[i\Delta_L\hat{L}_i^z,\hat{W}(t)\right]+\left[L_i^+,\left(\gamma_{lj}-i\chi_{ij}^{(l)}\right)\hat{W}(t)\hat{L}_j^-+\left(\gamma_{lr}-i\chi^{(lr)}\right)\hat{W}(t)\hat{\mathcal{R}}_i^-\right])$$
$$+\left[\hat{\mathcal{R}}_i^+,\gamma_r\hat{W}(t)\hat{\mathcal{R}}_i^-+\left(\gamma_{lr}-i\chi^{(lr)}\right)\hat{W}(t)\hat{L}_i^--i\Omega_r\hat{W}(t)\right]+H.c. \tag{12.3}$$

Here γ_l, γ_r and γ_{lr} are the photon losses from cavity during the cooperative excitation of bounded proteins. We require that the density matrix be expressed by the normal product $W(t_{stat})=\hat{F}\left(\hat{\mathcal{R}}^+,\hat{\mathcal{L}}^+\right)$. Let us consider that external bimodal field exits only first ring from the coupling system represented in the Fig. 12.2 so that $\Omega_r\gg\Omega_l$. In this particular case we obtain the stationary equation only for right filament/ring represented in Fig. 12.2:

$$\left[\hat{\mathcal{R}}_r^+,\left\{\gamma_r\hat{\mathcal{R}}_i^-+\left(\gamma_{lr}-i\chi^{(lr)}\right)L_l^--i\Omega_r\right\}\hat{F}*\left(\hat{\mathcal{R}}^-,\hat{\mathcal{L}}^-\right)\right]=0 \tag{12.4}$$

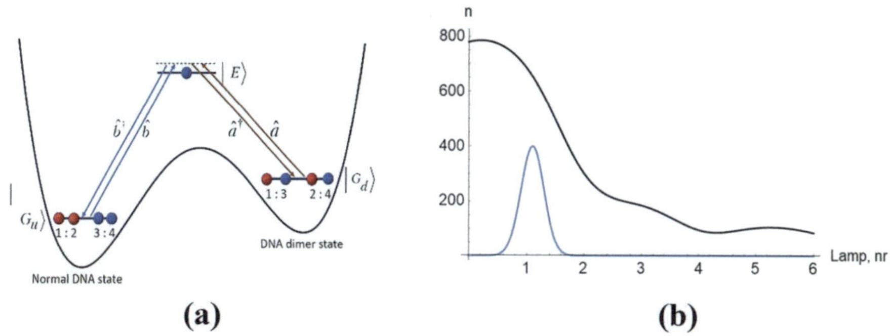

Fig. 12.3 Raman induced phase transition: **a** Description of the Hydrogen bonds between the nucleobases: Adenine and Thymine by two covalent Hydrogen bonds: A = T represents the normal state of the DNA molecule and dimer, T = T, is described by two covalent bonds between two thymines

Considering that for R filament/ring the interaction with L filament we neglect the term $\left(\gamma_{lr} - i\chi^{(lr)}\right)\hat{\mathcal{L}}_l^-$ from the Expression (12.4) in comparison with $i\Omega_r$. In this situation the solution is reduced to well known expression [17–19],

$$\hat{F} * \left(\hat{\mathcal{R}}^-, \hat{\mathcal{L}}^-\right) = \prod_{l=1}^{N} \frac{1}{-i\Omega_r + \gamma_r \hat{\mathcal{R}}_i^-} f\left(\hat{L}^-\right).$$

Here N is the number of right circuits in the possible packing procedures described in Figs. 12.3 and 12.4. Here we neglect the interaction between the right rings separated by left ones f the right ring molecules have only two states, than $\hat{\mathcal{R}}_i^{+2} = 0$, in this situation we have the identity, $\frac{1}{-i\Omega_r + \gamma_r \hat{\mathcal{R}}_i^-} = \frac{1}{-i\Omega_r} \exp\left(-i\frac{\gamma_r}{\Omega_r}\hat{\mathcal{R}}_i^-\right)$.

The solution for R ring becoms, $\hat{W}\left(\hat{\mathcal{R}}^+, \hat{\mathcal{R}}^-\right) = Aexp\left[+i\frac{\gamma_r}{\Omega_r}\sum_{l=1}^{N}\left(\hat{\mathcal{R}}_l^+\right)\right]\exp\left[-i\frac{\gamma_r}{\Omega_r}\sum_{m=1}^{N}\left(\hat{\mathcal{R}}_m^-\right)\right]$

From this expression follows that the rings passed into coherent state. This state induced in the left dipole–dipole "circuits" the same coherent state in which instead of external field we have polarization of right circuits. This effect induced the nonzero polarization and in the left ring/filament (see Fig. 12.4). The solution of this stationary equation is, described by the expression,

$$\hat{W}\left(\hat{L}^+, \hat{L}^-\right) = D\left(\frac{1}{\hat{L}^- + \chi^*\hat{\mathcal{R}}^+}\right)\left(\frac{1}{\hat{L}^- + \chi^*\hat{\mathcal{R}}^-}\right)$$

where, $\chi = (\gamma_{lr} - i\chi^{(lr)})/(\gamma - i\,\overline{\chi}^{(l)})$, D and A are normalization constants.

B. The similar situation may be in DNA where the base pairs in dipole–dipole interaction become more stable relative dipole active state of symmetric anbonded nuleobases [6–10, 20–23]. In other words, the dipole–dipole interaction between

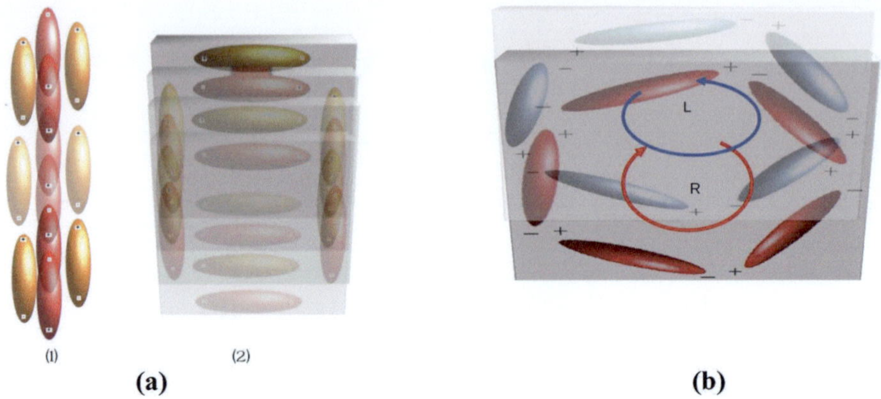

Fig. 12.4 Two types of spatial structures formed by tetramers **a** represented as H and J dipoledipole interaction according to Kasha model [25]. The first is made up of four polypeptide chains. The second of the? Sandwich as in the structures formed by trimmers. The formed multi-mers (described above, see pentamers in **b** may induce the left polarization in the new pentamer bio-molecule attracted new structure lake in Fig. 12.2. This coherent excitation and syntheses of new molecule may stay in the bases of construction of tubulin proteins represented

nucleobases hydrogen bonds can protect DNA from the active dipole excitation of solar radiation, making it immune to the new quantum correlation method. Indeed, taking into consideration the two opposite dipole of Hydrogen bonds between the Adenine and Timine (A: T), we observe that the ground degenerate state, I_{1a}, and I_{1b},, of two Hydrogen bonds may split into two according to perturbation theory. Denoting this two hydrogen bonds through the a and b we find the equation $(I_{1a} - E)((I_{2a} - E) - \mathrm{w}^2$,where in case when the Hydrogen antiparallel dipoles are perpendicular to distance between them we have the interaction energy, $W = 1b\left|\hat{H}_{1M}\right|1a \sim -d_a d_b/r^3$. Considering that the energy, I_{1a} and I_{1b}, slightly differ, $I_{1b} = I_{11} + \Delta$, we obtain the energies for ground state like in the case for multi-mers.

Let as firstly eliminate the atomic polarization, considering the special preparation of them in the initial coherent state. For this, considering that the lifetime of the atomic polarization between the excited and ground states of radiators is less than the inverse value of their Rabi nutation. The photon conversion will take place between two cooperative modes of the cavity represented in Fig. 12.5a, b is described by Hamiltonian (12.2) of the atomic system in multiple interaction with cavity field represented in Figs. 12.5. The rate of the generated number of atoms passing into ground state, $\partial D_z(t)/\partial t$, can be detected outside the cavity using ionizing technique. This detection method gives us the information about generation rate of the mean number of portion of energy, $\partial \hat{W}_{ph}(t)/\partial t$, emitted during the flaying time of the system of bomolecule components through the optical resonator. Considering that the quasi-particle operators, $\hat{\wedge}\ (t)$, as a function of $\hat{\wedge}_n^-(t)$, $\hat{\wedge}_n^+(t)$ and $W_{ph}(t) = \hat{\wedge}_n^z + L/2$, we can represent the atomic operators through the free, and sources parts as this is proposed in Refs. [14–16]. The conservation lows follows from the description of the system Hamiltonian. We observe that due to the multiple conversions of the total number of photons in the system, $\hat{N}_{ph} = \sum_{p=0}^{n} \hat{c}_p^\dagger \hat{c}_p$, and total number of atoms, $N_r = \sum_j \{|e\rangle\langle e|_j + |g\rangle\langle g|_j\}$,are conserving too. As the in the resonance case the free part of the Hamiltonian, $H_0 = \hbar\omega\hat{W}_{ph} + \hbar\omega\hat{D}_z + \hbar\omega N_{ph}$, commutes with interaction one, $\hat{H}_I$, we can introduce the conservation of the number of excitation portions of energy in the system, $\hat{W}_C(t) = \sum_{m=0}^{n} m\hat{c}_p^\dagger(t)\hat{c}_m(t) + \hat{D}_z + \frac{N}{2} = Const$. The last conservation low open the possibility to generate in the cavity energy portions multiple to $\hbar\omega$. If we wont to detect the number of the energy portions, $\hat{W}_{ph} = \sum_{m=0}^{n} m\hat{c}_p^\dagger(t)\hat{c}_m(t)$, generated in the cavity, we may use this operator in the free part of Hamiltonian, $H_0 = \hbar\omega\hat{W}_{ph} + \hbar\omega\hat{D}_z + \hbar\omega N_{ph}$. Here as in the tradition resonances interaction between the field and atomic subsystem we introduce the detuning from resonance, $\delta = \omega_r - \omega$, considering it relative smaller than radiator frequency, ω_r.

For unification of the two scattering conversions represented in Fig. 12.5, it is convenient to use the number of atoms prepared in coherent state, $|\phi_r(0)\rangle = \exp[\hat{D} - \eta]|j, j/[1 + |\eta|^2]^j$, where $\eta = \exp(i\phi)\tan(\theta/2)$, $|j, j$ is the excited state of the atomic ensemble passing through the cavity. Following the permutation procedure under the trace operation, in interaction picture we obtain the following master equation for these cavity quasi-particles described by density operator equation,

$$\frac{\partial}{\partial t}\check{W}(t) = (k + i\zeta)\left[\hat{\Lambda}_n^+ \check{W}(t), \hat{\Lambda}_n^-\right] + i\chi_n 2\eta\left[\check{W}(t), \hat{\Lambda}_n^-\right] + H.c. \qquad (12.5)$$

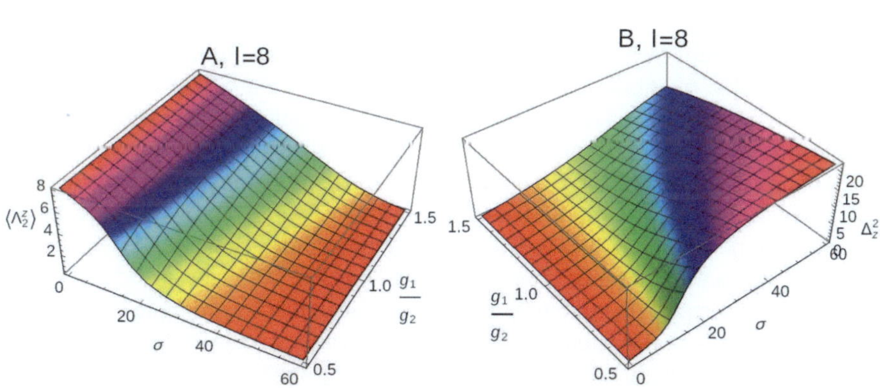

Fig. 12.5 **a** The system of bomoleculas is trapped into the coherent state near excited state, $l = 8$. This coherent system converts, $l = 8$, photons into the higher energy quanta of the discrete cavity states. Achieving the cooperative expectation the conversion process relax into the state with inversion near excited state, $\hat{\Lambda}_2^z - l$

Here $k = 2\,\chi_n^2\, j\gamma/(\delta^{\,2} + \gamma^2)$, is the per photon spontaneous conversion rate, and $\zeta = 2\,\chi_n^2\,\delta\, j/(\delta^{\,2} + \gamma^2)$ is the Raman shift from resonance stimulated by interaction with atomic ensemble. For the small value of polar angle, θ, we can neglect the higher order terms relative to the small parameter, $|\eta| < 1$.

For the small value of $|\eta|$, we can neglect the higher order terms so that for zero value of the detuning from resonance, $\delta = 0$, we obtain the following stationary equitation,$k\left[\hat{\wedge}_n^+ \overset{\vee}{W}(t) + i\sigma \overset{\vee}{W}(t), \hat{\wedge}_n^-\right] + H.c. = 0$, the stationary solution of which can be represented in the normal product form, relative to the operators, $\hat{\wedge}_n^+$ and $\hat{\wedge}_n^-, \overset{\vee}{W} = A\left[1/(\hat{\wedge}_n^+ + i\sigma)\right]\left[1/(\hat{\wedge}_n^- - i\sigma)\right]$. Here $\sigma = \Omega/(2k)$, A is the normalization constant of the density matrix of the quasi particle excitations. The behavior of the moments of the inversion operator and its quantum fluctuation are plotted in Fig. 12.5a–c for the initial number of photons, $l = 8$. As follows from this representation the fluctuations of the normal, $\Delta_{2pn}^2 = (\hat{\wedge}_2^+)^2(\hat{\wedge}_2^-)^2 - (\hat{\wedge}_2^+\hat{\wedge}_2^-)^2$ and anti-normal moments $\Delta_{2pa}^2 = (\hat{\wedge}_2^-)^2(\hat{\wedge}_2^+)^2 - (\hat{\wedge}_2^-\hat{\wedge}_2^+)^2$, may be estimated by the stationary solutions of master equations This square fluctuation, Δ_{2p}^2, achieved the minimal value in the region where relative conversion amplitudes between the first and second step $g_1/g_2 = 1.4$, coherent excitation, $\sigma = 40$.

12.3 Results and Discussion

The Raman model demonstrates that the dimerization rate of thymine bases ($T = T$) increases exponentially with the intensity of UVC radiation, suggesting the involvement of a quantum mechanism in dimer formation. This process has been experimentally confirmed through resonant fluorescence techniques, which highlight the disruption of hydrogen bonds and associated structural transitions. In our research, we observed that membrane proteins can be assembled via dipole–dipole interactions involving four, five, six, or more polypeptide chains, as confirmed through experimental evidence. Although the mathematical formulations underlying these observations are complex and are not detailed here, we identified a noteworthy rule: planar protein structures composed of an odd number of polypeptide chains are unable to form H- type hydrogen bonds. This packing procedures illustrated in Figs. 12.3 and 12.4 using a tetramer model as an example. Further, we investigated the spatial wrapping of these protein structures in three dimensions. One example involves the wrapping of equilateral triangular motifs, analyzed under π rotational symmetry relative to their diagonals. In these"sandwich-like" configurations, the dipoles exhibit a reduced electric moment compared to that observed within the plane of the structure. This model provides a foundation for applying perturbation theory to the energy states derived from V-type dipole wrapping. Additionally, we propose analogous structural models for tetramers depicted in Fig. 12.4, where the V -type dipole trapping mechanism plays a critical role in determining the stability and energetic properties of the protein assemblies. An innovative method of repackaging optical metamaterials using variable-sized quartz fibers is proposed for disinfection of pathogenic infected liquids with UVC radiation. The main concept is the rotation of the contaminated liquid through screw channels integrated in the metamaterial made of UVC spherical optical fibers, which improves the decontamination efficiency (see Fig. 12.5). To validate the approach, we conducted a study of dynamic and static inactivation of bakery yeast using UV regimes. In the static test, we used 80 ml of fungus solution, setting a short decontamination time. The yeast solution was prepared from 25 g of yeast dissolved in 800 ml of water and 20 g of agar ($C14H24O9$) to support the development of fungus. At the initial stage, we observed the formation of variable yeast colonies. Previous experiments have demonstrated their inactivation in 15 min, and in this phase we achieved inactivation after 24 h of growth. We investigated the impact of UVC light intensity on the inactivation rate, adding up to six UVC lamps to the decontamination core, according to Fig. 12.6.

12.4 Conclusions

The methods of repackaging proteins and biomolecules in supra-molecular structures using dipolar interactions and multiple photon conversions have been proposed. A quantum cooperation between photons and biomolecular subdivisions, within Raman scattering processes, allow us to obtain a formation and control of aggregate biomolecular like multi-meri. The study of quantum correlations opened possibilities for teleportation of information between the conversion stages and improving the efficiency of energy transfer. The application of these principles of generation and destroying of supramolecular structures near the surface of optical metamaterials by ultraviolet C radiation are proposed for improving the decontamination rate of pathogens inactivations.

Multiple dimerizations can involve photon correlations in Raman conversion, in which trains of coherent waves break weak bonds between nucleobases, contributing to the formation of multi-dimers. Another cooperative type can be, for example, the cooperative Dicke effect, which involves a collective transition between the normal and dimerized/multimerized states with long living time between adjacent nucleotides. Experimentally, other types of generation of multimeric biomolecules can be generated from tertiary monoacid chains described as subcomponents of protein structures under the action of radiation in multiple Raman scattering. This process can be observed in the generation of excitons of such a molecular suprastructures, but at the same time this mechanism can be used for the inactivation of pathogens, for example, using the dimerization/multimerisation of DNA or the protein substructures with other molecular aggregate from the living cells.

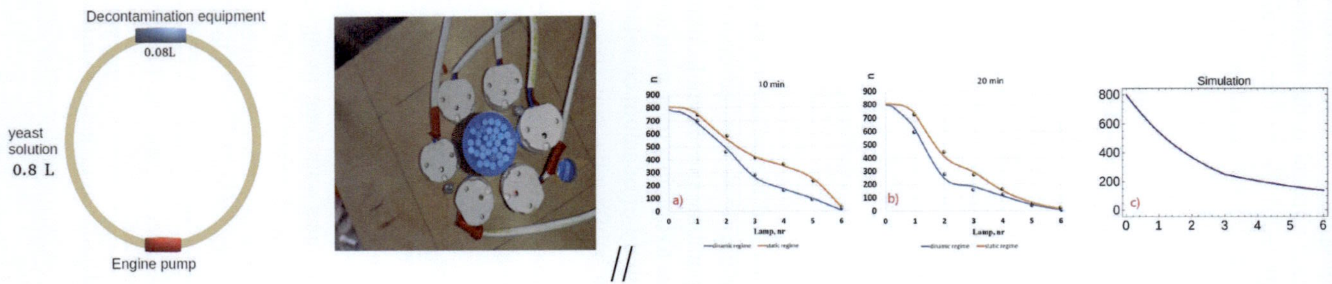

Fig. 12.6 Decontamination speed obtained with left-hand equipment according to the applied intensity of UVC radiation in units of number of lamps: **a** 10 Min dynamic and static regimes; **b** 20 Min dynamic and static regimes; **c** The dependence of the inactivation rate according to the intensity applied typically in accordance with the proposed theoretical model [24]

Acknowledgement The authors gratefully acknowledge the scientific and material support provided by the Institute Project "Quantum Cooperative Phenomena, 011206", the State Project "Stimulating Excellence" in Research for the Years 2024–2025, 24.80012.5007.26SE and 25.80012.5007.36TC.

References

1. Reece, J., Campbell, N.: Biology. Benjamin Cummings, San Francisco. ISBN 978-0-8053-6624-2 (2006)
2. Alberts, B.: Chromosomal DNA and its packaging in the chromatin fiber. In: Molecular biology of the cell, 4th ed., p 207. Garland Science, New York. ISBN 978-0-8153-4072-0 (2002)
3. Youngson, R.M.: Collins dictionary of human biology. HarperCollins, Glasgow. ISBN 978-0-00-722134-9 (2006)
4. Cox, M., Nelson, D.R., Lehninger, A.L.: Lehninger principles of biochemistry. W.H. Freeman, San Francisco. ISBN 978-0-7167-4339-2 (2005)
5. Redon, Pilch, D., Rogakou, E., Sedelnikova, O., Newrock, K., Bonner, W.: Histone H2A variants H2AX and H2AZ. Curr. Opin. Genet. Dev. **12**(2), 162–169. PMID 11893489.16 (2002). https://doi.org/10.1016/S0959-437X(02)00282-4
6. Wang, H., Zhang, L., Huang, J., Yang, Z., Fan, C., Yuan, L., Zhao, H., Zhang, Z., Liu, X.: Imaging the intracellular refractive index distribution (IRID) for dynamic label-free living colon cancer cells via circularly depolarization decay model (CDDM). Biomed. Opt. Express **15**(4), 2451–2465 (2024). https://doi.org/10.1364/BOE.518957
7. Chamot, S., Migacheva, E., Seydoux, O., Marquet, P., Depeursinge, C.: Physical interpretation of the phase function related parameter γ studied with a fractal distribution of spherical scatterers. Opt. Express **18**, 23664–23675 (2010). https://doi.org/10.1364/OE.18.023664
8. Brunsting, Mullaney, P.F.: Differential light-scattering from spherical mammaliancells. Biophys. J. **14**, 439–453 (1974)
9. Maier, J.S., Walker, S.A., Fantini, S., Franceschini, M.A., Gratton, E.: Possible correlation between bloodglucose concentration and the reduced scattering coefficient of tissues in the near-infrared". Opt. Lett. **19**, 2062–2064 (1994)
10. Liu, H., Beauvoit, B., Kimura, M., Chance, B.: Dependence of tissue optical properties on solute-induced changes in refractive index and osmolarity. J. Biomed. Opt. **1**, 200–211 (1996)
11. Blombergen, N., Shen, Y.R.: Quantum-theoretical comparison of nonlinear susceptibilities in parametric media, lasers, and Raman lasers. Phys. Rev. **133**, A37–A49 (1964)
12. Shen, Y.R. Blombergen, N.: Theory of stimulated Brillouin and Raman scattering. Phys. Rev. **137,** A1787-A1805 (1965)
13. Wang, C.-S.: Theory of stimulate Raman scattering. Phys. Rev. **182**, 482–494 (1969)
14. Enaki, N.A.: Cooperative properties of multiple quantum scattering: I quantum nutation. Phys Scripta. **99**(4), 045102 (2024). https://doi.org/10.1088/1402-4896/ad29cb
15. Enaki, N.A.: Cooperative properties of multiple quantum scattering: II. Coherentlasing. Phys Scripta. **99**(4), 045101 (2024). https://doi.org/10.1088/1402-4896/ad2ac0
16. Enaki, N.A.: Mutual cooperative effects between the mode components of two-photon and Raman induced cavity lasing processes. Opt. Commun. **498**, 127124 (2021). https://doi.org/10.1016/j.optcom.2021.127124
17. Puri, R.R., Lawande, S.V.: Exact steady-state density operator for a collective atomic system in an external field. Phys. Lett. A **72**, 200–202 (1979)
18. Puri, R.R., Lawande, S.V., Hassan, S.S.: Dispersion in the driven Dicke model. Optics Commun. **35**, 179–184 (1980)
19. Drummond, P.D., Walls, D.F.: Quantum theory of optical bistability. I: Nonlinear polarisability model. J. Phys. A **13**, 725 (1980). https://doi.org/10.1088/0305-4470/13/2/034

20. Kurian, P., Dunston, G., Lindesay, J.: How quantum entanglement in DNA synchronizes doublestrand breakage by type II restriction endonucleases. J. Theor. Biol. **21**(391), 102–112 (2016). https://doi.org/10.1016/j.jtbi.2015.11.018
21. Mukherjee, S., Majumdar, S., Bhattacharyya, D.: Role of hydrogen bonds in protein-DNA recognition: effect of nonplanar amino groups. J. Phys. Chem. B **109**(20), 10484–10492 (2005). https://doi.org/10.1021/jp0446231
22. Singh, V., Fedeles, B.I., Essigmann, J.M.: Role of tautomerism in RNA biochemistry. RNA **21**(1), 1–13 (2015). https://doi.org/10.1261/rna.048371.114
23. Slocombe, L., Sacchi, M., Al-Khalili, J.: An open quantum systems approach to proton tunnelling in DNA. Commun. Phys. **5**, 109 (2022). https://doi.org/10.1038/s42005-022-00881-8
24. Munteanu, I., Starodub, E., Bazgan, S., Turcan, M., Paslari, T., Podoleanu, D., Enaki, N.A.: Ultraviolet C intensity dependence of decontamination efficiency for pathogens as function of repacked metamaterials with screw channels. Eur. Biophys. J. **53**(3), 133–145 (2024). https://doi.org/10.1007/s00249-024-01702-2
25. Khan, U., Kasha, M.: Mechanism of four-level laser action in solution excimer and excitedstate proton-transfer cases. Proc. Natd Acad. Sci. USA **80**, 1767–1770 (1983)

Chapter 13
Qualitative Analysis of Some Benzodiazepines Performed by Densitometry

Ana-Maria Hossu, Ionica Ionita, Aurora Anca Poinescu, Daniela Avram, Alexandru Stoica, and Cosmina Cristiana Avram

Abstract The addictive potential of benzodiazepines and the need for strict dosage control led to numerous studies on the qualitative and quantitative determination of these. The possibility of qualitative determination of diazepam, oxazepam, nitrazepam and clonazepam in the mixture using semi-automated sputtering, manual development, UV detection and densitometry was investigated.

Keywords Benzodiazepines · HPTLC · Densitometry

13.1 Introduction

Benzodiazepines are a class of drugs widely used in psychiatry, targeting the central nervous system. Thus, benzodiazepines are successfully used as sedatives, anxiolytics, hypnotics, muscle relaxants, anti-convulsant, in the treatment of alcoholism or as pre-medication in other medical (surgical) or dental treatments.

Although they were introduced into medical practice by Hoffman-La-Roche in the 1960s, their adverse effects, the most important of which is their addictive potential, only became apparent in the 1980s.

A.-M. Hossu (✉) · I. Ionita
Department of Sciences and Advanced Technologies, Valahia University of Targoviste, Targoviste, Romania
e-mail: ana.hossu@valahia.ro

A. A. Poinescu
Department of Material Engineering, Valahia University of Targoviste, Targoviste, Romania

D. Avram · A. Stoica
Department of Food Engineering, Valahia University of Targoviste, Targoviste, Romania

C. C. Avram
SC Almina Trading SA, Targoviste, Targoviste, Romania

R. Tamas and C. A. Heiman (eds.), *Selected Papers from AToMS 2024*,
Lecture Notes on Multidisciplinary Industrial Engineering,
https://doi.org/10.1007/978-3-032-02452-7_13

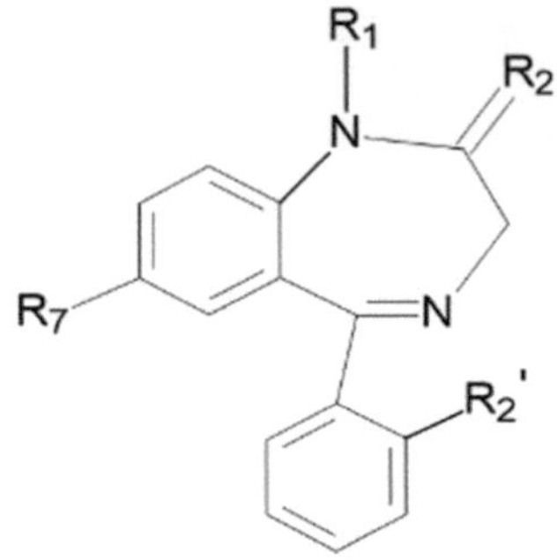

Fig. 13.1 General structure of benzodiazepines

This reason, as well as the need for strict dosage control, led to numerous studies on the qualitative and quantitative determination of benzodiazepines, all the more so because of their particular structure, consisting of a benzene nucleus fused with a heterocycle of 7 atoms, of which 2 nitrogen atoms in positions 1 and 4 (Fig. 13.1).

Table 13.1 presents a systematized selection of articles [1–8] for determination of benzodiazepines by TLC.

13.2 Materials and Methods

Table 13.2 shows the coding of the materials used. All dilutions were carried out in methanol. As stationary phase, silica gel 60 GF_{254} deposited on 20 × 20 cm glass plates (200 μm layer thickness, fluorescence indicator for 254 nm), automatically traced, purchased and used as such from Merck (without prior pre-activation) was used.

The following solvents and substances were used for the mobile phases: acetone (Chimopar S.A.), ammonia 25% (Chimactiv S.A.), chloroform (Chimopar S.A.), n-hexane (Merck), methanol (HPLC, Sigma), NaBr p.a. (Chimopar S.A.), $NaHCO_3$ (Merck), NaOH (Merck) 6 M, in distilled water.

Plates were strip-sputtered under nitrogen jet using a semi-automatic plate-sputtering system from Camag (Switzerland), model Linomat 5 and analyzed by densitometry using a chromatoplate scanning system from Camag (Switzerland) model TLC Scanner 3, software WinCats manager ver. 1.4.4. Initially, a spouting plan of the substances of interest was made, containing: the solutions spooled, the order of their placement on the plate, the volumes to be spooled for each migration lane.

Initially, a spouting plan of the substances of interest was made, containing: the solutions spooled, the order of their placement on the plate, the volumes to be spooled for each migration lane.

In accordance with the chromatoplate plan, well-defined volumes of the solution to be analyzed were sputtered on each plate. The plates were developed in the dark, in a vertical development chamber, previously saturated with mobile phase vapor.

Table 13.1 Comparative analysis of some data from the related literature to the study of some benzodiazepines by TLC

The composition of the samples	Chromatographic systems: Stationary phase Mobile phase	Reference
Screening from a mixture of diazepam, desmethyldiazepam, temazepam, oxazepam, ketazolam, oxazolam	Silica gel Toluene Detection after spraying with Bratton-Marshall reactive and şi diazotizing with $NaNO_2 + HCl$	[1]
Diazepam, nordiazepam	Silica gel Chloroform: ethyl acetate (3:1)	[2]
Diazepam, oxazepam, chlordiazepoxide	Silica gel Chloroform: methanol (9:1)	[3]
Oxazepam (7 intermediates of all oxazepam synthesis steps)	Silica gel Chloroform: ethyl acetate (3:1)	[4]
Oxazepam, diazepam, nitrazepam, chlordiazepoxide, medazepam	RP-18 Methanol: acetate buffer pH = 3 (3:1) Detection after spraying with ethanol solution 2% nitrobenzene and, after drying, with ethanol solution 10% NaOH, or by exposure to vapors of HCl, or spraying with Dragendorff reactive	[5]
Medazepam, nitrazepam, chlordiazepoxide, oxazepam	Silica gel Chloroform: ether: isopropanol (84:13:3)	[6]
1,4-benzodiazepine	Silica gel Chloroform: acetone 4:1; benzene Derivatization with cu Bratton-Marshall reactive	[7]
Mixture of 6 benzodiazepines	Silica gel n-hexane: acetone (6:4)	[8]

Table 13.2 Coding of used materials (used later in tables and figures)

Code	Substance
D	Diazepam p.a. (Sigma-Aldrich), stock solution of 1 mg/mL in HPLC methanol
O	Oxazepam p.a. (Merck), stock solution of 1 mg/mL in HPLC methanol
N	Nitrazepam p.a. (Sigma-Aldrich), stock solution of 1 mg/mL in HPLC methanol
C	Clonazepam p.a. (Sigma-Aldrich), stock solution of 1 mg/mL in HPLC methanol
F	Flunitrazepam p.a. (Sigma-Aldrich), stock solution of 1 mg/mL in HPLC methanol
M	Midazolam maleate p.a. (Sigma-Aldrich), stock solution of 1 mg/mL in HPLC methanol
Me	Medazepam p.a. (Sigma-Aldrich), stock solution of 1 mg/mL in HPLC methanol
Am1	O + D + N + C (1:1:1:1)
Am2	M + F + C (1:1:1)

Table 13.3 Mobile phases used for densitometric analysis of mixtures of benzodiazepines for which good separations were obtained

System	Mobile phase	Spiked substances
1	Methanol: ammonia (40:0.6)	D, Me, N, O
2	n-hexane: methanol: acetone (15:3:2)*	D, C, N, O, M, Me
3	Chloroform: acetone (8:2)	D, M, O
4	n-hexane:acetone (3:2)	D, O, N, C, M, F

* The plate was previously washed with methanol

After development, the plates were left in the dark to evaporate the mobile phase and then evaluated by densitometry, based on the Rf values provided by the software and the UV spectra of the migrated spots. Table 13.3 shows the systems that led to the best results, and Table 13.4—the Rf values obtained for each of these mobile phases.

From the analysis of the results (shown in Tables 13.3 and 13.4), as well as from the chromatoplate analysis, it can be seen that M.p. 1 separates diazepam and oxazepam from midazolam relatively well, but the first two benzodiazepines co-elute; M.p. 2 generally separates diazepam and oxazepam well from the other benzodiazepines, but nitrazepam and flunitrazepam can co-elute; M.p. 3 does not discern diazepam from midazolam; M.p. 4 leads to the best separation values of the 6 benzodiazepines tested.

It follows from this that the analytical procedure development step is of utmost importance for establishing the working parameters for the development of the validated method.

13.3 Method Validation

A. *Analytical procedure*

Validation is applied to the method for the TLC analysis of diazepam, oxazepam, nitrazepam and clonazepam in the mixture using semi-automated sputtering, manual development, UV detection and densitometry.

The procedure for the preparation of samples and chromatoplates shall be described in as much detail as possible.

In the present situation, semi-automated strip sputtering was carried out under nitrogen jet on F254 silica gel plate on 20×20 cm glass support (Merck), followed by manual development in vertical plane with n-hexane: acetone mixture (3:2, v/v), detection by densitometry at 310 nm.

B. *Selectivity (specificity)*

Figure 13.2 shows a chromatoplate spiked with D, O, N, C and their mixture (Am1) and Table 13.5—software generated ratio for this plate.

Table 13.4 Results obtained for Rf values for analysis using the mobile phases in Table 13.3

M.p	D	O	N	C	M	F	Me
1	0.14 ± 0.01	0.15 ± 0.01	–	–	0.24 ± 0.01	–	–
2	0.61 ± 0.02	0.36 ± 0.01	0.47 ± 0.01	0.44 ± 0.02	0.07 ± 0.03	0.48 ± 0.01	
3	0.61 ± 0.01	0.18 ± 0.01	–	–	0.61 ± 0.01	–	0.12 ± 0.03
4	0.61 ± 0.02	0.36 ± 0.01	0.48 ± 0. 0.01	0.32 ± 0.01	0.07 ± 0.01	0.39 ± 0.02	–

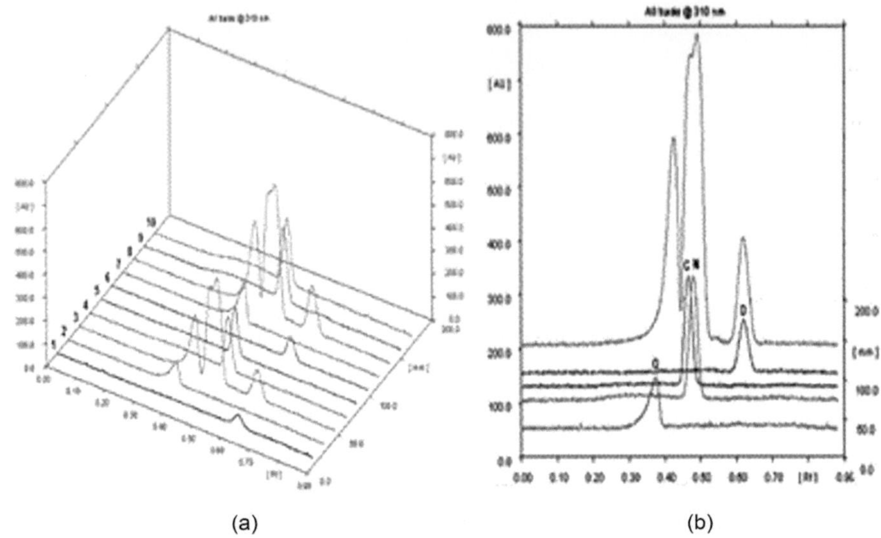

Fig. 13.2 (**a**) Spot chromatoplate with 2 μL D, O, N, C (lanes 1, 2, 4 and 5, respectively), 3 μL D, O, N, C (lanes 6, 7, 9 and 10, respectively), 4 μL Am1 (lane 3) and 8 μL Am1 (lane 8); **b** selection of lanes to highlight the separation of the four bezodiazepines in the mixture

The absorption spectra plotted for each migrated spot can also be used to validate the identification (such an example is shown in Fig. 13.3). Also from Fig. 13.3 it can be seen that although the four benzodiazepines have absorption at 254 nm (so can be detected in the UV), the optimal wavelength for their detection is around 310 nm (where all four have high absorbance).

C. *Accuracy and precision*

In the case of densitometric analysis, accuracy and precision can be assessed based on data from tables such as Table 13.5, since the software generates very accurate values of the Rf parameter (the "Max position" column), on the basis of which averages, and standard deviation can be calculated for each migrated spot.

D. *Linearity range*

Of course, in the case of identification analysis it is not necessary to establish a linearity domain. However, in the validation of the method, it is good to specify the minimum and maximum concentrations that can be tested, as well as the minimum and maximum volumes of spots deposited on the start line. Both parameters may influence the detection of one of the compounds in the mixture if it is in too high a concentration and/or deposited in too large a volume, relative to another compound with a similar Rf value that is in a low concentration—in which case co-eluting may occur.

Table 13.5 Software-generated report for chromatoplate analysis shown in Fig. 13.2a

Track	Peak	Start Position	Start Height	Max Position	Max Height	Max %	End Position	End Height	Area	Area %	Assigned substance
1	1	0.60 Rf	18.5 AU	0.63 Rf	57.6 AU	100.00	0.66 Rf	14.3 AU	6107.6 AU	100.00	diazepam
2	1	0.32 Rf	15.8 AU	0.37 Rf	95.6 AU	100.00	0.39 Rf	8.3 AU	10,589.9 AU	100.00	oxazepam
3	1	0.36 Rf	45.5 AU	0.40 Rf	256.3 AU	20.58	0.42 Rf	13.0 AU	26,295.8 AU	24.58	oxazepam
3	2	0.43 Rf	17.5 AU	0.46 Rf	408.3 AU	32.77	0.46 Rf	401.8 AU	27,925.5 AU	26.11	clonazepam
3	3	0.47 Rf	402.1 AU	0.48 Rf	453.1 AU	36.38	0.52 Rf	9.5 AU	39,865.3 AU	37.27	nitrazepam
3	4	0.59 Rf	27.4 AU	0.62 Rf	127.9 AU	10.27	0.65 Rf	13.7 AU	12,882.8 AU	12.04	diazepam
4	1	0.45 Rf	12.8 AU	0.48 Rf	230.9 AU	100.00	0.51 Rf	6.6 AU	19,151.5 AU	100.00	nitrazepam
5	1	0.44 Rf	10.3 AU	0.47 Rf	204.7 AU	100.00	0.49 Rf	10.5 AU	16,203.4 AU	100.00	clonazepam
6	1	0.59 Rf	8.8 AU	0.62 Rf	98.2 AU	100.00	0.65 Rf	9.2 AU	9239.8 AU	100.00	diazepam
7	1	0.39 Rf	56.5 AU	0.42 Rf	173.4 AU	100.00	0.44 Rf	8.4 AU	15,953.7 AU	100.00	oxazepam
8	1	0.38 Rf	59.2 AU	0.43 Rf	385.5 AU	22.66	0.44 Rf	65.9 AU	41,413.0 AU	27.05	oxazepam
8	2	0.44 Rf	66.0 AU	0.47 Rf	539.8 AU	31.73	0.48 Rf	538.0 AU	40,019.0 AU	26.14	clonazepam
8	3	0.48 Rf	537.7 AU	0.49 Rf	576.4 AU	33.88	0.53 Rf	15.1 AU	51,026.0 AU	33.32	nitrazepam
8	4	0.59 Rf	12.9 AU	0.62 Rf	199.6 AU	11.73	0.65 Rf	13.3 AU	20,664.0 AU	13.50	diazepam
9	1	0.45 Rf	24.8 AU	0.48 Rf	327.2 AU	100.00	0.51 Rf	14.4 AU	29,587.4 AU	100.00	nitrazepam
10	1	0.43 Rf	31.3 AU	0.46 Rf	294.1 AU	100.00	0.49 Rf	18.9 AU	26,982.3 AU	100.00	clonazepam

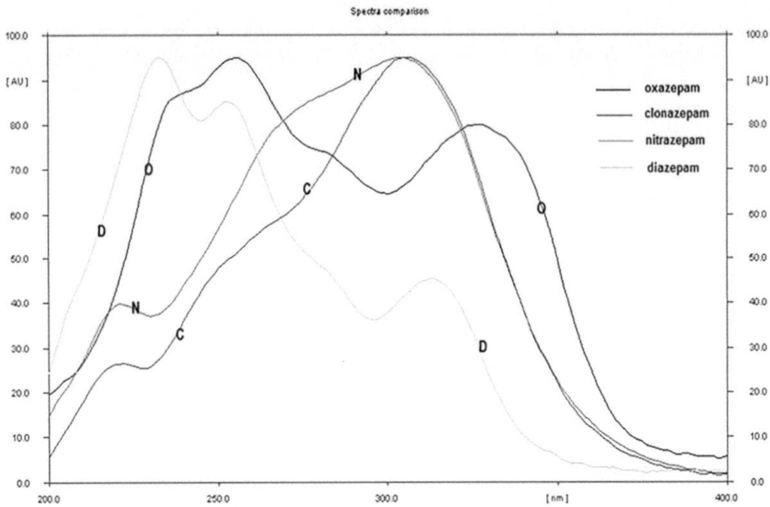

Fig. 13.3 Spectral comparison of chromatographic spots obtained for clonazepam, diazepam, oxazepam and nitrazepam

13.4 Conclusions

The possibility of qualitative determination of diazepam, oxazepam, nitrazepam and clonazepam in the mixture using semi-automated sputtering, manual development, UV detection and densitometry was investigated.

For this purpose, chromatoplates were made on silica gel 60 GF_{254} deposited on glass plates as stationary phase, and 4 variants were tested as mobile phase, the 4th system leading to the best separation values of the 6 benzodiazepines tested.

Semiautomatic stripping of the plates in strips, under nitrogen jet, with stock methanolic solutions of benzodiazepines was performed using the Linomat 5 device from Camag (Switzerland). For the evaluation of the plates after mobile phase development we used a TLC Scanner3 densitometer from Camag (Switzerland), assisted by WinCATS software ver 1.4.4.

Validation parameters determined were selectivity, accuracy and precision, and linearity range.

References

1. Schuetz, H., Fitz, H., Suphacherabhan, S.: Screening und Nachweis von Ketazolam und Oxazolam, Arzneim. Forsch **33**, 507–512 (1983)
2. White, D.J., Stewart, J.T., Honigberg, I.I.: Quantitative analysis of diazepam and related compounds in drug substance and tablet dosage from HPTLC and scanning densitometry. J. Planar Chromatogr. **4**, 413–415 (1991)

3. Sarbu, C., Cîmpan, G.: Determination of 1,4-benzodiazepines by quantitative TLC. A regression study employing the ladder of power. J. Planar Chromatogr. **9**, 126–128 (1996)
4. Hopirtean, I., Bota, I., Hopirtean, I., Marutoiu, C.: J. Planar Chromatogr. **12**, 392–394 (1996)
5. Kastner, P., Klimes, J.: Analysis of benzodiazepines by adsorption and ion-pair RP TLC, J. Planar Chromatogr. **1996**(9), 385–387 (1997)
6. Cimpoiu, C., Hodisan, T., Nascu, H.: Comparative study of mobile phase optimisation for the separation of some 1,4-benzodiazepines. J. Planar Chromatogr. **10**, 195–199 (1997)
7. Volf, K.: Detection of 1, 4-benzodiazepines and benzophenones with a secondary amino group by thermal conversion and diazotation and spraying with Bratton-Marshall reagent. J. Planar Chromatogr. **11**, 132–136 (1998)
8. Sprinteroiu, M., Balalau, D., Guţu, C.M., Blendea, L., Ilie, M.: Studii CSS pentru determinarea calitativă şi semicantitativă a unor benzodiazepine din amestec, publicata în volumul simpozionului Noi provocari in practica farmaceutica, Oradea, pp. 278–283. ISBN 978-973-759-891-2 (2009)

Chapter 14
Analytical Models of Experimental Data Expressed as Diagrams in the Automatic Materials' Selection

Emil M. Oanţă

Abstract The paper presents original approaches based on numeric solutions, which replace the analytic methods used in the solvers of the structural problems, in this way being accessed a higher level of generality and being reached an improved accuracy of the results. Firstly, in the materials' fatigue evaluation, the safety domain is redefined in an original way, in order to avoid the imprecision due to the linearization of their frontiers, which would lead to smaller coefficients of safety. Another original aspect consists of the spline interpolation of the diagrams that store the information regarding the factors, which influence the fatigue limit, in this way the materials' fatigue calculi being automatically performed. The application is used to evaluate several materials and to provide the safety coefficients that become, together with other materials' constants, input data in a multi-criteria-decision-making (MCDM) software. In this way, the resulted hierarchy of materials assist the structural analyst to select an optimal material. Several original software instruments for research are developed in Python and GNU Octave.

Keywords Automatic CAD · Analytic to numeric models transition · Materials' fatigue automatic calculus · MCDM · Optimization

14.1 Introduction

Key features of the classic approaches in engineering were based on simple models in-line with the assumptions on top of which the analytical theory was developed, similar to the mathematical theories based on axioms. This approach was widely used in the 'early' age when the grounds of engineering were conceived, which was a historic era of science development with elementary means to perform calculi, being desired relevant results using a minimal volume of not-so-sophisticated calculations.

E. M. Oanţă (✉)
Department of General Engineering Sciences, Constanta Maritime University, Constanţa, Romania
e-mail: emil.oanta@universitateamaritima.ro

R. Tamas and C. A. Heiman (eds.), *Selected Papers from AToMS 2024*,
Lecture Notes on Multidisciplinary Industrial Engineering,
https://doi.org/10.1007/978-3-032-02452-7_14

Accordingly, simplifying hypotheses, aka assumptions, were extensively used. For practical reasons information was expressed graphically, the diagrams being easily to be interpreted by the experts and appropriate values to be extracted from diagrams and used in the calculus simple methodologies. Extensive experimental work was performed in order to have statistically confirmed reliable data, which were used to calibrate the theoretical models, the according diagrams being still useful nowadays in their classic, graphical form. The invaluable contribution of the 'forefathers' of the classic analytical approaches is far too valuable not to be reconsidered in the actual intelligent computing context. Beside various and distinct branches of science which inspired the aforementioned remarks, we may focus onto the materials' fatigue specific issues because variable loads seem to be ubiquitous in structural problems to be solved in many areas of engineering.

14.2 State of the Art

One can notice the nowadays wide spread of the valuable software instruments created by collectives of bright intelligences around the world. A trend in the last decades was the integration of complementary software instruments, i.e. design, simulation, manufacturing, document management and project management, in a unique package in order to offer complete computer based solutions of a given class of problems. Another trend was to include application program interface libraries in the commercial software for engineering, fact that open the possibility to include them as 'building blocks' in complex research projects where the data 'flows' between the 'data processing units'. An additional trend was to include in the commercial software for engineering new facilities that increase the effectiveness of the analysts, such as: prototyping, parametric design that read the values of the dimensions from neutral format data files, i.e. CSV, JSON, which may be generated by distinct outer applications. All these trends are the foundation for novelty and highly creative methods of investigation, which integrate several solving concepts into hybrid methodologies of research. The according results of the integrated methods are useful in several ways: to be compared against the solutions resulted from other (types of) studies or to be considered as initialization of the solutions to be found in more complex research iterative methods.

14.3 Original Solution

The purpose of the study is to create data processing units useful in materials' fatigue studies, reusable in other research projects, based on a mathematical general background and implemented in a flexible way. Original approaches of this research may be remarked at several levels and addressing several problems. In this section is presented an overview regarding the relevant aspects of the original approaches, fact that makes easy to understand the novelty of the research strategy. In the following sections are presented the computing aspects and the according results and details regarding the innovative aspects.

A. *Conceiving a more accurate safety domain model*

The flexible definition of the calculus domains over which complex calculi may be performed is a long run research topic, original mathematical and computing solutions being conceived by us. In this case study, the safety domain is drawn in the (σ_m, σ_A) coordinates; $\sigma_m = \frac{\sigma_{max}+\sigma_{min}}{2}$ is the mean stress, $\sigma_A = \frac{\sigma_{max}-\sigma_{min}}{2}$ is the stress amplitude and $R = \frac{\sigma_{min}}{\sigma_{max}}$ is the stress ratio; σ_{min} and σ_{max} are the extreme values of the stresses in a stationary loading; σ_{-1} is the endurance limit, or the fatigue limit and it represents the maximum stress of the fully reversed cycles, $R = -1$, for which the specimen withstands variable loads for an indefinite number of cycles without failing; σ_0 is the endurance limit of an "one to max" cycle for which either $\sigma_{min} = 0$, or $\sigma_{max} = 0$, i.e. $R \in \{0, \infty\}$; the fatigue endurance is expressed by $\psi = \frac{2\cdot\sigma_{-1}-\sigma_0}{\sigma_0}$; we denote by σ_Y the yield stress and by σ_r the ultimate tensile strength according to the stress–strain curves, [1]. The stresses specific to the materials fatigue phenomenon either may result from experiments, or may be deduced using as references the stresses in the stress–strain diagram, [2–4]. As a test study we consider the Haigh model, the Soderberg model and a new model of the safety domain designated as the 'modified Haigh' domain, as they are presented in Fig. 14.1; the following reasoning may be easily applied to another models of the safety domain. The points in the diagram have the following coordinates: $A(0, \sigma_{-1})$, $B\left(\frac{\sigma_0}{2}, \frac{\sigma_0}{2}\right)$, $C(\sigma_r, 0)$ and $D(\sigma_Y, 0)$. Point E is located on the ABC curved frontier and its abscissa is σ_Y, therefore DE is parallel to the vertical axis.

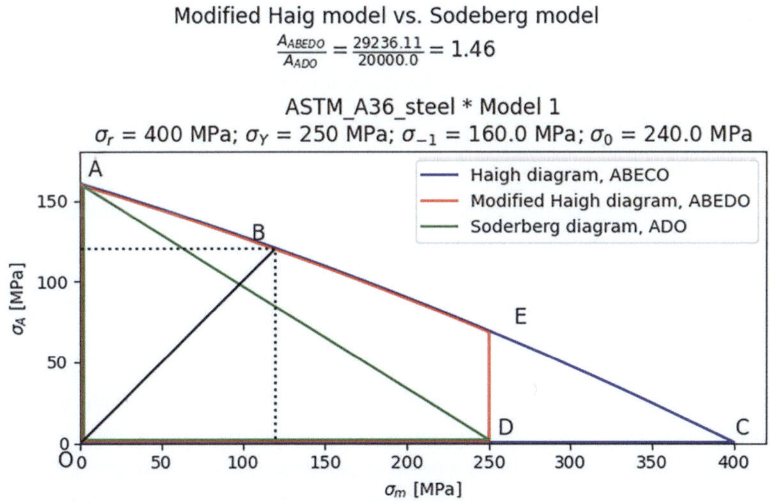

Fig. 14.1 Comparison between the models of the safety domains for two models of evaluated material fatigue characteristics of the same steel

The simplest analytical calculations consider the Soderberg model of the safety domain, i.e. the *ADO* triangle, for which the mean stress in static conditions is the yield stress (point *D*), assumption which is appropriate in the context of the mechanical engineering loadings. One can easily notice that the area of the *ABEDO* domain is larger than the area of the Soderberg domain, i.e. *ADO*. Working with the Soderberg model we get smaller safety coefficients, i.e. the structure is oversized and/or over-engineered. Therefore, it is reasonable to consider the *ABEDO* domain, designated as the modified Haigh model.

The *ABEC* curved frontier must be analytically defined in order to have an analytical definition of the modified Haigh model, that is important for computing purposes. A first approach is based on the remark that the *A*, *B* and *C* points may be used to deduce a second degree polynomial which passes through these points. This solution was applied using both 'pencil-and-paper' method and also two symbolic solvers, [5, 6]. The resulting function of the 'modified Haigh' diagram frontier for which we use the notations: $x \leftarrow \sigma_m$, $y \leftarrow \sigma_A$ is:

$$f_{mH}(x, y) = y - a_2 \cdot x^2 - a_1 \cdot x - a_0 = 0, \tag{14.1a}$$

$$y = a_2 \cdot x^2 + a_1 \cdot x + a_0 \tag{14.1b}$$

$$\begin{cases} a_2 = \frac{2\cdot\sigma_{-1}\cdot\sigma_0+2\cdot\sigma_r\cdot\sigma_0-4\cdot\sigma_r\cdot\sigma_{-1}}{2\cdot\sigma_r^2\cdot\sigma_0-\sigma_r\cdot\sigma_0^2} \\ a_1 = \frac{\sigma_{-1}\cdot\sigma_0^2+2\cdot\sigma_r^2\cdot\sigma_0-4\cdot\sigma_r^2\cdot\sigma_{-1}}{2\cdot\sigma_r^2\cdot\sigma_0-\sigma_r\cdot\sigma_0^2} \\ a_0 = \sigma_{-1} \end{cases} \tag{14.2}$$

The function of the *ABEC* curved frontier may be also deduced using spline functions, same as they are used for the factors that influence the endurance limit fatigue, or using some other types of functions, [7].

Starting from the function (14.1a), one can notice that $f_{mH}(x, y)$ may have a general form, i.e. a quadratic curve or it may be a simple equation of a line. In this general context, let us consider a randomly defined stationary loading cycle define by the extreme values $\sigma_{\min C}$ and $\sigma_{\max C}$ for which are calculated σ_{mC} and σ_{AC} stresses. If the $(\sigma_{mC}, \sigma_{AC})$ point is located inside the *ABEDO* region, the structure does not fail under the variable loads. This remark is analytically expressed by the conditions

$$\begin{cases} \sigma_{mC} \le \sigma_Y \\ f_{mH}(\sigma_{mC}, \sigma_{AC}) \cdot f_{mH}(0, 0) \ge 0 \end{cases} \tag{14.3}$$

The first condition is limiting the stress in static conditions up to the yield stress. The second condition states that the $(\sigma_{mC}, \sigma_{AC})$ point must be on the same side in respect to the *ABE* frontier, same as a point of reference, in this case being considered the (0, 0) point. The result is zero if the $(\sigma_{mC}, \sigma_{AC})$ point is located on the *ABE* frontier. This qualitative condition offers a binary answer. However, a quantitative aspect based on the method of the allowable quantities may use the condition

$$\sigma_{maxC} \le \sigma_{mC} + \sigma_{AABE} \tag{14.4}$$

where σ_{AABE} is the ordinate resulted from the equation $f_{mH}(x = \sigma_{mC}, y = \sigma_{AABE}) = 0$. Moreover, a factor of safety may be deduced for this particular case:

$$c = \frac{\sigma_{mC} + \sigma_{AABE}}{\sigma_{maxC}} \tag{14.5}$$

Further considerations and developments may use the analytical version of any safety domain model and the factor of safety previously deduced. Condition (14.5) may be also used in an iterative selection of a material for which $c \to c_a$, where the allowable value may be, to the limit, $c_a \to 1$.

B. *Computer aided analytic model of the curves that express the factors that influence the fatigue limit*

Considering the factors which modify the fatigue limit, the Soderberg safety coefficient is

$$c = \frac{1}{\frac{K_f}{\varepsilon \cdot \gamma} \cdot \frac{\sigma_A}{\sigma_{-1}} + \frac{\sigma_m}{\sigma_r}},\ c = \frac{1}{\frac{K_f}{\varepsilon \cdot \gamma} \cdot \frac{\sigma_A}{\sigma_{-1}} + \frac{\sigma_m}{\sigma_Y}}, \tag{14.6}$$

while the Seresen coefficient is

$$c = \frac{1}{\frac{K_f}{\varepsilon \cdot \gamma} \cdot \frac{\sigma_A}{\sigma_{-1}} + \psi \cdot \frac{\sigma_m}{\sigma_r}} \tag{14.7}$$

K_f is the fatigue stress concentration factor, defined as:

$$K_f = 1 + \eta_k \cdot (\alpha - 1) \tag{14.8}$$

for which η_k is the material's fatigue notch sensitivity and α is the stress concentration factor. ε is the size factor and γ is surface finish factor. The α, η_k, ε, ε_k and γ factors are mostly expressed as diagrams resulted from experimental studies, being also available approximate formulae, such as Neuber's empirical formula and Peterson's empirical formula, that were used in valuable research studies, [8, 9]. The best option is to use statistical validated experimental data expressed as diagrams and employed in a plethora of research projects, being an important source of safe and relevant information. Diagrams expressed analytically may be used in automatic dimensioning, in this way being used interpolation methods, such as spline interpolation. This method uses third degree polynomials for the subdomains of a curve, being necessary to compute the according coefficients. A first stage is to collect (x_i, y_i), $i = 1, n$ points along the curve. The next stage is to use this input data to compute the a_{ki} coefficients,

$$s(x) = \sum_{i=1}^{n-1} H_{i,i+1}(x_i, x_{i+1}) \cdot \underbrace{\left[\sum_{k=0}^{3} a_{ki} \cdot (x - x_i)^k\right]}_{f_{Si,i+1}(x)}, \tag{14.9a}$$

$$H_{i,i+1}(x_i, x_{i+1}) = \left[H(x - x_i) - H(x - x_{i+1})\right] \tag{14.9b}$$

where $H(x - x_i)$ is the Heaviside step function, $H_{i,i+1}(x_i, x_{i+1}) = \left[H(x - x_i) - H(x - x_{i+1})\right]$ is 1 inside the x_i, x_{i+1} interval and 0 outside, and $f_{Si,i+1}(x)$ is the spline function in the $i, i + 1$ interval.

This method is implemented in GNU Octave and it generates a CSV file having the structure $(\#i, x_i, x_{i+1}, a_{3i}, a_{2i}, a_{1i}, a_{0i})$. It also generates source code in GNU Octave, Java and C + + which output: (a) n the number of points; (b) the a_{ki} coefficient for some given k and i; (c) $s(x)$ for a given x. In the automatic generated libraries, the spline data are hard-coded, therefore the code may be very fast executed and the libraries may be included as modules embedded in specialized software applications, fact which enhances the flexibility of the solution.

Using as input data the diagrams presented in [2] there were converted to analytical form several graphs of the α, η_k, ε, ε_k and γ factors which modify the endurance limit, some of them being presented in Figs. 14.2 and 14.3, in this way being now possible to analytically compute K_f using (14.8), as well as the safety coefficients, using (14.6) and (14.7).

This analytical method is used for the automatic dimensioning, the output dimensions being written in CSV files which are read by the CAD (parametric design) files in order to set the current values of the parameters, e.g. in NX.

C. *Aspects regarding the MCDM method*

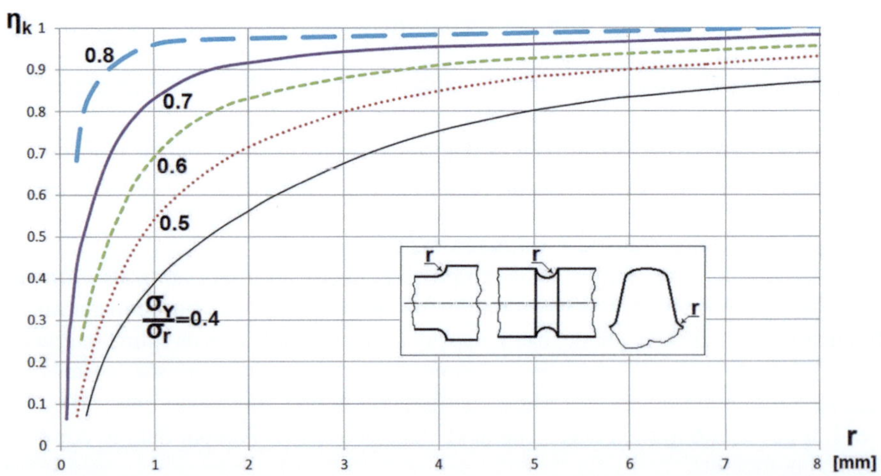

Fig. 14.2 Notch sensitivity factor for various $\frac{\sigma_Y}{\sigma_r}$ ratios

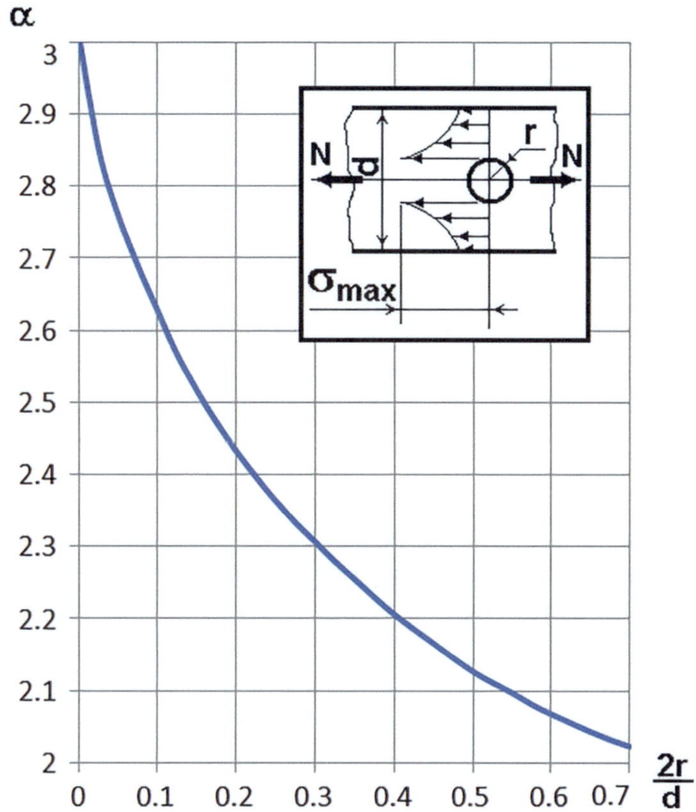

Fig. 14.3 Variation of the stress concentration factor with respect to the $\frac{2r}{d}$ geometric dimensions

Making decisions is an omnipresent process in science, an optimal design being the result of a succession of smart decisions. In this study the selection of a material whose properties best fit the functional requirements is based on the integration the aforementioned novelty aspects and uses the Multiple-Criteria-Decision-Making method.

Table 14.1 presents a decision making problem having m variants of action, $(V_1, V_2, \cdots, V_m)$, n criteria, $(C_1, C_2, \cdots, C_n)$, and a state of nature, $k = 1$ because we have a certainty scenario, while, in general problems, there are considered r states of the nature, $(S_1, S_2, \cdots, S_r)$. In this case study the V_i variants are the designations of the materials, the C_j criteria are the designations of the material constants relevant for the context of the current decision making, and $x_{i,j}$ are the values of the materials' constants. If a criterion is optimized by maximization, such as the fatigue strength, the according utility is $u(V_i) = \frac{x_i - x_{min}}{x_{max} - x_{min}}$, while the utility of a criterion optimized by minimization, such as the price, the utility is $u(V_i) = \frac{x_{max} - x_i}{x_{max} - x_{min}}$. In this way, the matrix of the consequences is transformed in the matrix of the utilities. The general utility of the i variant of action is computed using

Table 14.1 Matrix of the consequences in a certainty conditions scenario

Variants	Criteria S_k			
	C_1	C_2	...	C_n
V_1	$(x_{1,1})_{S_k}$	$(x_{1,2})_{S_k}$	...	$(x_{1,n})_{S_k}$
V_2	$(x_{2,1})_{S_k}$	$(x_{2,2})_{S_k}$	...	$(x_{2,n})_{S_k}$
⋮	⋮	⋮	⋮	⋮
V_m	$(x_{m,1})_{S_k}$	$(x_{m,2})_{S_k}$	...	$(x_{m,n})_{S_k}$

$$u_{gli} = \sum_{j=1}^{m} (\gamma_j \cdot u_{ij}) \tag{14.10}$$

and the optimal variant is the one for which (10) has a maximum value, this method being designated the weighted sum model. γ_j is the relative weight of importance of the j criterion and they are computed using the analytic hierarchy process, AHP, [10].

The weighted product method uses the level of performance,

$$H_{absi} = H_{Tabs} \cdot \left[\prod_{j \in S_{\max}} \left(\frac{x_{ij}}{x_{refj}} \right)^{\gamma_j} \right] \cdot \left[\prod_{j \in S_{\min}} \left(\frac{x_{refj}}{x_{ij}} \right)^{\gamma_j} \right] \tag{14.11}$$

where x_{refj} is the j consequence of the variant of action selected as reference, for which $H_{Tabs} = 1000$, S_{max} is the optimisation by maximisation set of criteria and S_{min} is the optimisation by minimisation set of criteria. The relative level of performance is

$$H_{reli} = \frac{H_{absi}}{max_l (H_{absl})} \tag{14.12}$$

and the optimal variant is the one for which (12) has a maximum value.

An example of a hierarchy of materials resulted using the weighted product method is given in Table 14.2.

14.4 Computing Aspects and Results

The novelty approach regarding the safety domains' models offers more realistic results, taking into account the entire domain, not only a subdomain with a linear frontier. The ratio between the area of the new domain and the initial linear-frontier area is $\frac{A_{ABEDO}}{A_{ADO}} = 1.46 \cdots 1.68$ which is relevant for the increase of the new domain.

Table 14.2 Weighted product method hierarchy

Initial index	Variants of action	WPM hierarchy index	Level of performance
6	Steel A514, HAZ, BHN = 461	0	1503.50
5	Steel A514, BHN = 303	1	1157.87
7	Steel RQC-100, Su = 863.0	2	1059.58
1	Steel A27, Cast, BHN = 135	3	1000
0	Steel A-517 Grade F, BHN = 256	4	936.75
3	Steel A36, HAZ, BHN = 243	5	904.55
4	Steel A36, Su = 540.0	6	756.87
2	Steel A36, BHN = 160	7	584.31

The calculi were performed using symbolic solvers, together with an original software developed in Python consisting of 550 computer code lines, whose results are presented in Fig. 14.1. The ideas may be used to conceive new safety domain computer aided models.

The factors which modify the fatigue limit, expressed as curves and the frontier of the safety domains' models were analytically modelled using an original GNU Octave software. The output CSV files were customized in Excel in order to emphasize the relevancy of the visual information in its scientific particular context, Figs. 14.2 and 14.3. The original software generates source code in GNU Octave, Java and C++, which can be integrate in meta-level applications. A library of analytical models was created and any new project will enrich the library with new curves which will be also converted to an analytical form.

The decisional problem was solved by developing a software instrument in Python consisting of more than 1400 computer code lines. For flexibility and reusability reasons, each problem was solved in a distinct class. The classes are linked one to the other using CSV files. An updated version is planned which will use data frames 'flowing' between the classes. It is expected to use the software instruments in a wide range of projects, testing their output against the machine learning results.

14.5 Conclusion

Original software instruments development is a rational research strategy, which integrates knowledge acquired along eras of scientific evolution and nowadays-powerful computers in the context of the increased use of the artificial intelligence and machine learning in all fields of science. The paper presents ideas, methods and the according implementations, which solve materials' fatigue issues, that may be used in further studies meant to create new safety domain models, and specific dimensioning software for computer aided design, a work frame being presented in Fig. 14.4.

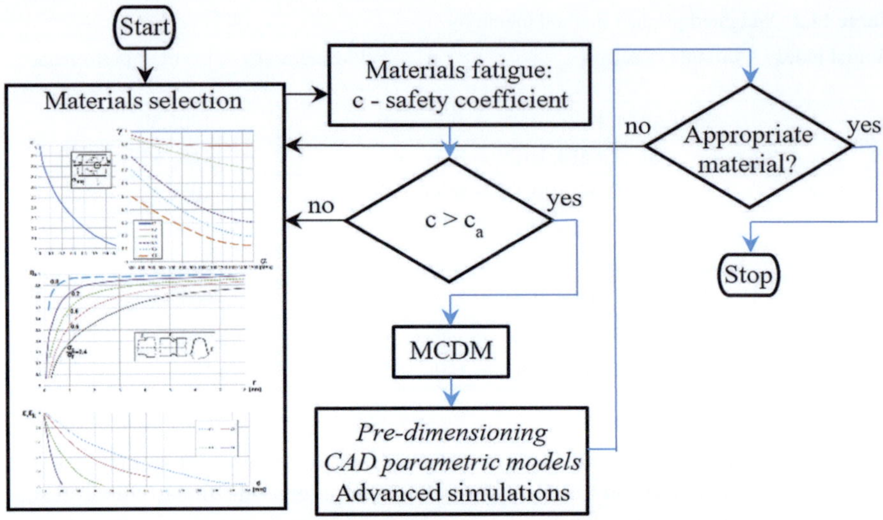

Fig. 14.4 Flowchart using the original models previously presented

Selection of the appropriate materials in the early stages of a project is important to avoid loops, which would increase the demand of additional resources. The decisional methods and implementations may be used in a plethora of investigation topics and science projects in inter-domains research.

References

1. Stress–strain curve. https://en.wikipedia.org/wiki/Stress%E2%80%93strain_curve. Accessed 15 June 2024
2. Buzdugan, G.: Strength of materials. 'Technical' Publishing House, Bucharest (1979)
3. Retno-Anggraini, T., et al.: Tensile strength/yield strength (TS/YS) ratios of high-strength steel (HSS) reinforcing bars. In: AIP Conference Proceedings of ISMM2017, Jakarta (2018) https://doi.org/10.1063/1.5038318
4. Bossard Knowledgebase: What is the difference between tensile strength and yield strength? https://provenproductivity.com/what-is-the-difference-between-tensile-strength-and-yield-strength/. Accessed 15 April 2024
5. Wolfram Alpha. https://www.wolframalpha.com/input?i=systems+of+3+equations+calculator. Accessed 15 June 2024
6. QuickMath. https://quickmath.com/. Accessed 15 June 2024
7. Chui, C.K.: Wavelet analysis and its applications. In: An introduction to wavelets, vol. 1. Academic Press (1992)
8. Yadegari, P., et al.: Approximation of elastic-plastic local strains in surface-hardened notched components. In: Abstracts of MATEC Web of Conference 349, ICEAF-VI 2021 04008 (2021). https://doi.org/10.1051/matecconf/202134904008
9. Luo, P., Yao, W., Ssusmel, L., Li, P.: Prediction of fatigue damage region with the use of the notch critical plane approach for crack initiation and propagation. Int. J. Fatigue 135–105533 (2020). https://doi.org/10.1016/j.ijfatigue.2020.105533

10. Zumrut, I.B., Baran, H.B., Ozbalta, T.G.: Multi-criteria decision-making approach for selecting a structural system of an industrial facility. OTMC **14**(1), 2656–2665 (2022). https://doi.org/10.2478/otmcj-2022-0010

Part III
Applications in Mechanical and Industrial Engineering

Chapter 15
Thermodynamic Analysis of a Current Marine Refrigeration System in the Case of Sea Water Cooled Condenser

Feiza Memet

Abstract On board the ships, vapour compression technologies are the most suitable ones. Seen the significant amount of energy used for cooling perishables during voyages, energy efficiency gains more importance. In present times, marine refrigeration is dominated by HFCs, a key role being played by R134a. Despite that R134a is a potent greenhouse gas, marine refrigeration systems using this refrigerant are subject of energy consumption improvement. In this paper it is performed a first and second laws based analysis under the influence of sea water temperature. This analysis is very important in the case of the direct use of sea water as a coolant in refrigeration condensers on board—especially in the context of global warming. It is investigated the effect of condensation temperature (sea water temperature) on the Coefficient of Performance, work input, heat rejection in the condenser, total exergy destruction and exergy efficiency. One of the most significant findings to emerge from this study is that high sea water temperatures accelerate work consumption and total exergy destruction - with negative impact on energy efficiency of the system.

Keywords Marine refrigeration · Energy efficiency · Sea water temperature

15.1 Introduction

Maritime transportation acts as a catalyst for the international trade [1], energy efficiency in ships being a way to boost competitiveness [2]. Marine refrigeration is an application of refrigeration systems used to preserve refrigerated or frozen perishables in optimal conditions of temperature and air humidity and also to add comfort on board—for crew and passengers (on passenger ships). There is a strong connection between the global economic increase and the intensification of refrigerated cargo transport; this trend is seen in the last decades and continues to respond to the

F. Memet (✉)
General Sciences Department, Constanta Maritime University, Constanta, Romania
e-mail: feizamemet@yahoo.com

R. Tamas and C. A. Heiman (eds.), *Selected Papers from AToMS 2024*,
Lecture Notes on Multidisciplinary Industrial Engineering,
https://doi.org/10.1007/978-3-032-02452-7_15

growing demand throughout specialized reefer carrier fleets and reefer containers [3]. In present times, the sustainable transport of perishable goods is accepted as a goal focusing on decarbonization and energy efficiency increase [4, 5]. Reference [4], highlights that the International Maritime Organization (IMO) works to decrease emissions from ships by at least 40% by 2023 compared with 2008 levels, one of measures being speed limit—a real challenge for marine refrigeration when coping both with safe delivery and environmental protection issues. On the other hand, as in [5], refrigerated cargo transport is linked to important energy consumption, the investigations on energy consumption decrease being needed for an efficient way to transport perishables.

Marine refrigeration plants are approached in a particular manner in comparison to the civil ones. Vapour compression technologies are the most suitable for on board operation, being seen safety, weight or dimensions aspects; on the other hand, cooling demand and fuel storage should be in good correlation, especially for voyages in extreme marine conditions [6]. As one of the most common used, in this refrigeration system the refrigerant is vaporized and condensed alternately in order to remove heat from a low temperature sink to a high temperature one. Refrigeration systems are cyclic devices operating between two pressure limits, with the refrigerant flowing in four states: compression, condensation, expansion and evaporation [7, 8].

In present times, marine refrigeration is dominated by HFCs, such as R134a, R404A or R407C. The mentioned refrigerants have a negative impact on the environment and are subject of the Montreal Protocol and Kyoto Protocol; in this respect, the "optimal" refrigerant should have characteristics such as: null Ozone Depletion Potential (ODP), low Global Warming Potential (GWP), no toxicity and flammability and excellent thermodynamic and thermophysical properties [9–11]. High global warming potential refrigerants are still not prohibited in marine refrigeration and the use of HFCs is not restricted, only in some cases, marine registers require a maximum value for GWP of around 2000 [12]. This is why more than 20% of merchant ships have in use R134a—a safe refrigerant, with null ODP but high GWP (1430) and with good thermodynamic properties.

Despite the fact that an important part of marine refrigeration plants use a potent greenhouse gas such as R134a, responsible refrigeration concept is aligned with the considerable energy consumption decrease and with energy efficiency achievement when R134a is in use as well. In this respect, this paper deals with the theoretical evaluation of energy and exergy performances of R134a in vapor compression refrigeration systems on board the ships. This theoretical study targets cargo ships with a cooling capacity of maximum 96 kW. The analysis is devoted to the influence of the sea water temperature on the condensing temperature, when the heat removed from the condenser in rejected into the sea water. Nevertheless, it is worth to pay attention to the direct use of sea water in condensers since the effect of sea water temperature increase is assumed to influence the performance of the plant, especially when this increase is linked to global warming [13]. Parameters like Coefficient of Performance (COP), compressor power, heat rejection in condenser, exergy efficiency and total exergy destruction are computed in this work.

15.2 Methods and Materials

Figure 15.1 illustrates T-s diagram with super heating and sub cooling. Are shown the main four processes: 1–2 isentropic compression, 2–3 condensation with heat rejection to the sea water, 3–4 isenthalpic throttling, 4–1 evaporation with heat absorption from the body to be cooled.

The following assumptions have been taken:

•all processes are steady state,

•pressure loses in the pipelines are neglected,

•heat gains to the system are neglected,

•heat losses from the system are neglected,

•kinetic, potential and chemical exergy are neglected,

•compressor's efficiency is 80%,

•degree of sub cooling of the refrigerant in liquid state is 5 °C,

•degree of super heating of vapors leaving the evaporator is 5 °C,

•evaporation temperature is –15 °C,

•condensation temperature varies in the range 31—41 °C.

The above mentioned range for the condensation temperature results from Eq. (15.1) which describes the connection between sea water conditions and condensation temperature [14].

$$t_c = t_{sw} + 13.5\ ^{\circ}\mathrm{C} \tag{15.1}$$

where: t_c—condensation temperature

t_{sw}—sea water temperature.

The specific of thermodynamic processes encountered in refrigeration systems is that these occur with important amounts of heat to be evacuated in the environment and the heat exchange between the system and the surroundings takes place at a

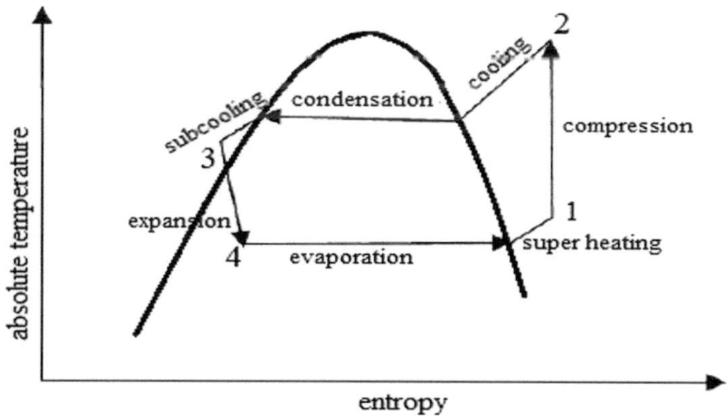

Fig. 15.1 Single stage vapor compression refrigeration cycle in T-s diagram

finite temperature difference. This fact leads to irreversibilities on the cycle which causes performance degradation. For these reasons, thermodynamic analysis of the refrigeration systems should be developed on first and second laws basis. Concerned on the energy conservation, the first law of thermodynamics allows the assessment of COP (Coefficient of Performance), which means the refrigeration effect per unit compressor work and represents an important factor for performance analysis. Due to the fact that processes met in refrigeration systems are irreversible, their accurate analysis should imply the exergy concept.

Exergy is referred to as the maximum obtainable work output from a system. System performance is enhanced when the exergy loss is at low levels. Exergy analysis is a strong tool in optimizing refrigeration systems and it is used to assess exergy destructions and exergy efficiency of the system, or second law efficiency. Exergy efficiency is seen as a more powerful performance indicator than COP and it is defined as the rate between output exergy and input exergy. Above were mentioned the advantages of this analysis which also has limitations. Considering economic concerns, the exergo- economic analysis is very useful. The exergo- economic analysis is very useful since it is able to establish the connection exergy- economics. While the proposed thermodynamic analysis deals with the energy conversion, thermo-economic analysis provides alternatives for cost- effective improvements. The equations specific to energy and exergy analysis conducted to assess the performance evaluation are provided below [15–20].

The performance of the system determined in term of the rate between the cooling capacity and the work required:

$$COP = (h_1 - h_4)/(h_2 - h_1) \tag{15.2}$$

where: h—specific enthalpy, kJ/kg.

The isentropic compressor work rate, in kW, need to increase the pressure of the refrigerant is calculated as follows:

$$\dot{W}_{comp} = \dot{m}_f(h_2 - h_1) \tag{15.3}$$

where: $\dot{m}_f$—refrigerant mass flow rate, kg/s.

The refrigeration effect, in kW, that defines the amount of cooling produced is calculated as follows:

$$\dot{Q}_{evap} = \dot{m}_f(h_1 - h_4) \tag{15.4}$$

The heat rate, in kW, rejected in the condenser is the energy removed from the refrigerant during condensing process and it is calculated as:

$$\dot{Q}_c = \dot{m}_f(h_3 - h_2) \tag{15.5}$$

The specific exergy in any state is written as:

$$ex = (h - h_0) - T_0(s - s_0) \tag{15.6}$$

where: s—specific entropy, kJ/(kgK), h_0 and s_0—specific enthalpy and specific entropy of the fluid, at the standard environment conditions T_0, p_0 ($T_0 = 298$ K, $T_0 = 1$ bar).

The rate of exergy destruction, in kW, is found as:

$$\dot{E}x_{Des} = \sum_j \left(1 - \frac{T_0}{T_j}\right)\dot{Q}_j - \dot{W}_{in} + \sum_i \dot{m}_{fi} e\,x_i - \sum_e \dot{m}_{fe} e\,x_e \tag{15.7}$$

where: $\dot{Q}_j$—heat transfer rate through the boundary, at absolute temperature "T_j" and at location "j".

Given the fact that friction, heat transfer or pressure drops in refrigerant lines are irreversible processes, the assessment of exergy destructions is very important in the thermodynamic analysis of the performance of refrigeration systems. The exergy destruction rate is equal with the irreversibility rate (I); thus:

$$\dot{E}x_{Des} = 1 \tag{15.8}$$

The total irreversibility (total exergy destruction) of the refrigeration system is given by the sum of the irreversibility (exergy destruction) of each component of the system. For each of the different components of the system are written the following equations:

For the evaporator:

$$\dot{E}x_{des,evap} = \dot{E}x_4 - \dot{E}x_1 + \dot{Q}_{evap}\left(1 - \frac{T_0}{T_e}\right) \tag{15.9}$$

where: T_e—evaporation temperature/(K).

$$I_{evap} = \dot{m}_f[(h_1 - h_4) - T_0(s_4 - s_1)] + \dot{Q}_{evap}\left(1 - \frac{T_0}{T_e}\right) \tag{15.10}$$

For the compressor:

$$\dot{E}x_{des,comp} = \dot{E}x_1 - \dot{E}x_2 + \dot{W}_{comp} \tag{15.11}$$

$$I_{comp} = \dot{m}_f[(h_1 - h_2) - T_0(s_1 - s_2)] + \dot{W}_{comp} \tag{15.12}$$

For the condenser:

$$\dot{E}x_{des,c} = \dot{E}x_2 - \dot{E}x_3 + \dot{Q}_c\left(1 - \frac{T_0}{T_c}\right) \tag{15.13}$$

where: T_c— condensation temperature/(K).

$$I_c = \dot{m}_f[(h_2 - h_3) - T_0(s_2 - s_3)] + \dot{Q}_c\left(1 - \frac{T_0}{T_c}\right) \tag{15.14}$$

For the expansion valve:

$$\dot{E}x_{des,Ev} = \dot{E}x_3 - \dot{E}x_4 \tag{15.15}$$

$$I_{Ev} = \dot{m}_f(s_2 - s_4)(since h_3 = h_4) \tag{15.16}$$

$$I_T = \dot{E}x_{des,T} = I_{evap} + I_{comp} + I_c + I_{Ev} \tag{15.17}$$

The exergy efficiency is calculated by:

$$\eta_{ex} = \dot{E}x_{out}/\left(\dot{E}x_{in} + W_{comp}\right) = 1 - \left(I_t/W_{comp}\right) \tag{15.18}$$

15.3 Results and Discussion

According to the above provided equations for the refrigerant cycle analysis, written by means of mass and energy conservation and after the determination of the thermodynamic properties for points 1, 2, 3, 4 of the cycle, Figs. 15.2, 15.3 and 15.4 show Coefficient of Performance, specific compressor work input and specific heat rejection in condenser values according to variable sea water temperatures, or, more precisely, according to variable condensation temperatures.

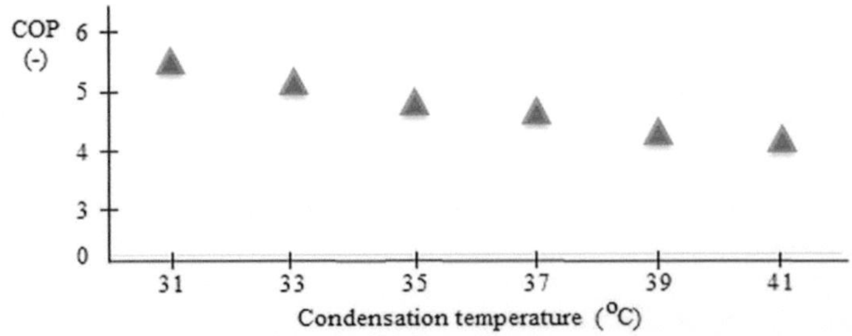

Fig. 15.2 COP in terms of variable condensation temperature

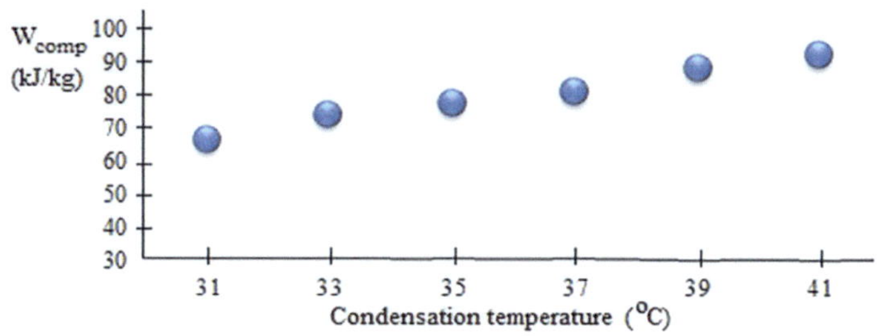

Fig. 15.3 A *Wcomp* in terms of variable condensation temperature

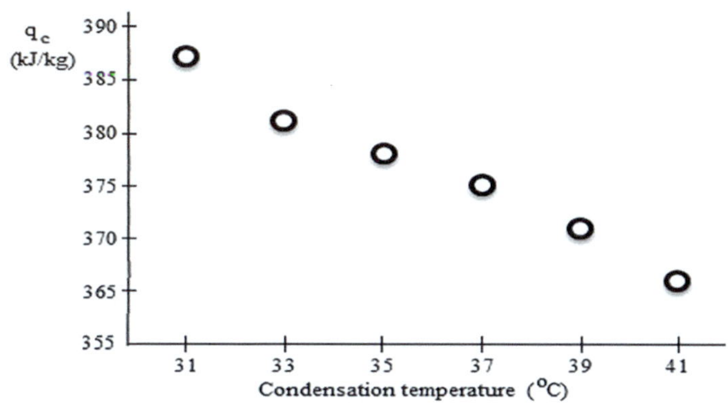

Fig. 15.4 A *qc* in terms of variable condensation temperature

As it can be seen from Fig. 15.2, for high condensation temperatures (high sea water temperatures) the Coefficient of Performance decreases. This decrease in the first law efficiency is explained by cooling capacity decrease and compressor power increase.

By keeping the evaporation temperature constant and increasing the condensation temperature, the compressor power consumption rises since increases the pressure ratio of the compressor (as seen in Fig. 15.3).

Because COP is inversely proportional to the compressor power consumption, from Fig. 15.2 was noticed the reduction in the performance of the cycle—in terms of COP. The condenser is an important component of the refrigeration system, when talking about system efficiency enhancement. As reveled in Fig. 15.4, the heat rejection in the condenser is slower when increases the temperature of the condenser cooling medium. The decrease of the condensing temperature is limited by sea water conditions.

In the case of high sea water temperatures, a good option is the increase of the effective heat rejection surface of the condenser. Based on the second law of thermodynamics are determined the total exergy destruction (I_T) and the exergy efficiency (ηex) of the analyzed refrigeration cycle. There is a direct proportionality between exergy destruction and the entropy generation—which is associated with the measurement of the irreversibilities which occur during the refrigeration cycle.

Less entropy generation means less exergy destruction and an improved exergy efficiency—a parameter which offers a better understanding of performance than COP; exergy efficiency delivers a realistic measure for performance enhancement. In Figs. 15.5 and 15.6 are plotted total exergy destruction and exergy efficiency variations with "t_c" varying from 31 °C to 41 °C.

High condensation temperatures, dictated by high sea water temperatures lead to the increase of the total irreversibility of the system.

As an effect, the exergy destruction increases—with a negative impact on the exergy efficiency values. Work consumption and total exergy destruction should be as minimum as possible. As it is seen, the increase in the condensation temperature do not permit the achievement of such conditions.

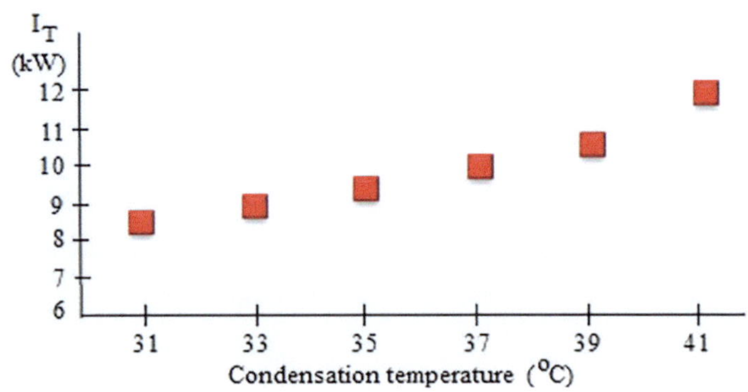

Fig. 15.5 Total exergy destruction versus condensation temperature

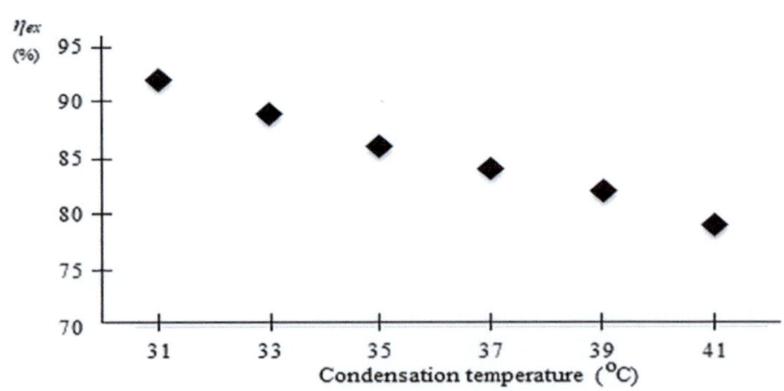

Fig. 15.6 Exergy efficiency versus condensation temperature

15.4 Conclusions

This study is initiated because marine refrigeration is known as an application that consumes large amounts of electrical energy and might use fluorinated gases. R134a is not prohibited from ships and, for many ship owners, retrofitting R134a is a reasonable cost only if the refrigeration system is a subject of major repairs. For this reason, energy use in actual vapour compression refrigeration systems using R134a is a must. In this paper, results related to a single stage vapour compression refrigeration were presented and discussed in terms of energy consumption. The system is designed for sea water direct cooling of the refrigeration condenser—a challenging situation when talking about the cooling medium.

Based on the laws of thermodynamics, the analysis was devoted to identify operational conditions allowing efficient energy consumption. The analysis was carried out at different condensation / sea water temperatures (31–41 °C), by keeping constant the evaporation temperature (−15 °C).

The results have demonstrated that the increase in condensation temperature directly affects the performance of the system. The Coefficient of Performance and exergy efficiency decreased with 25% and 14% respectively, because of increases in compressor work input (35%) and in total exergy destruction (42%). Also, the increase in the condensation temperature is found in a lower capacity of the condenser (7% decrease).

Acknowledgements Feiza Memet thanks to Constanta Maritime University for financing the publising of this article from its research funds.

References

1. Ion, A., Ticu R.I.: Dynamic analysis for calculating the marine structural design by using the finite element method. Constanta Marit. Univ. Ann. **22**(XV), 19–22 (2014)
2. Faitar, C., Nova, I.: Basic aspects and contributionsto the optimizationof energy systems exploitation of a super tanker ship. IOP Conf. Series: Mater. Sci. Eng. **227**(012043), 1–11 (2017)
3. Kotowska, I.: Transformations in organising maritime transport of perishable goods. Ekon. Probl. Uslug **124**, 7–16 (2016)
4. Zis, T.P.V., Psaraftis, H.N.: Impacts of short-term measures to decarbonize maritime transport on perishable cargoes. Marit. Econ. Logist. **24**(3), 602–629 (2022)
5. Filina-Dawidowicz, L., Filin, S., Wojnicz, L., Milek, D., Grzelak, P.: Energy efficient maritime transport of refrigerated containers. Procedia Comput. Sci. **207**(C), 3572–3581 (2022)
6. Sarish, B., Chealathan, M.: Energy conservation on-board ships thrugh enhancing energy management. Int. J. Recent Technol. Eng. **7**(6S6), 25–27 (2019)
7. Aziz, A., Mainil, A.K.: Performance comparison of refrigeration system using R134a and hydrocarbon refrigerant (HCR134a) with different expansion devices. IOP Conf. Series: Mater. Sci. Eng. **237**(1), 1–29 (2017)
8. Satsangi, R., Dheeraj, A., Kumar, S., Kumar, A., Singh, H., Satsangi, N.: R134a refrigerant in vapour compression cycle: a review paper. Int. Res. J. Eng. Technol. **05**(03), 507–511 (2018)

9. Tsvetkov, O.B., Laptev, Y.A., Sharkov, A.V., Mitropov, V.V., Feodorov, A.V.: Alternative refrigerants with low global warming potential for refrigeration and air-conditioning industries. IOP Conf. Ser.: Mater. Sci. Eng. **905**, 1–4 (2020)
10. Mansouri R., Bisulandu, B.J.R.M., Ilinca, A.: Assesing energy performance and environmental impact of low GWP vapor compression chilled water systems. Energies **16**(12), 1–16 (2023)
11. Aized, T., Rashid, M., Riaz, F., Hamza, A., Nabi, H.Z., Sultan, M., Ashraf, W.M., Krzywanski, J.: Energy and exergy analysis of vapor compression refrigeration system with low GWP refrigerants. Energies **15**(19), 1–22 (2022)
12. Wajs, J., Mrozek, M., Fornalik-Wajs, E.: Combined cold supply system for ship application based on low GWP refrigerants-thermo-economic and ecological analysis. Energy Convers. Manag. **258**(21), 1–16 (2022)
13. Teguh, N.H., Yuliati, L., Darmadi, D.B.: Effect of seawater temperature rising to the performance of Northern Gorontalo small scale power plant. Case Stud. Therm. Eng. 31–14 (2022)
14. Pigani, L., Boscolo, M., Pagan, N.: Marine refrigeration plants for passenger ships: low-GWP refrigerants and strategies to reduce environmental impact. Int. J. Refrig. INT. J. REFRIG. **64**, 80–92 (2016)
15. Altinkaynak, M., Olgun, E., Sencan Sahin, A.: Comparative evaluation of energy and exergy performances of R22 and its alternative R407C, R410A and R448A refrigerants in vapor compression refrigeration systems. El-Cezeri Int. J. Sci. Eng. **6**(3), 659–667 (2019)
16. Chandrasekharan, M.: Exergy analysis of vapor compression refrigeration system using R12 and R134a as refrigerants. Int. J. Students Res. Technol. Manag. **2**(04), 134–139 (2014)
17. Roy. R., Mandal, B. K.: First law and second law analysis of mechanical vapour compression refrigeration system using refrigerants CFC12, R134a and R290. Int. J. Curr. Eng. Tecnol. **3**(Special Issue), 191–195 (2014)
18. Baskaran, A., Matthews, K.: Exergetic analysis of a vapour compression refrigeration system with R134A, RE170, R429A, R435A. Int. J. Curr. Adv. Res. **6**(6), 4029–4036 (2017)
19. Aktaş, M., Dalkilic, A.S., Celen, A., Cebi, A., Mahian, O., Wongwises, S.: A theoretical comparative study on nanorefrigerant performance in a single-stage vapor-compression refrigeration cycle. Adv. Mech. Eng. **7**(1), 1–12 (2015)
20. Başhan, V., Kokkulunk, G.: Exergoeconomic and air emission analysis for marine refrigeration with waste heat recovery system: a case study. J. Mar. Eng. Techol. **19**(3), 147–160 (2019)

Chapter 16
Energy Efficiency Assessment of a Vapour Compression Refrigeration System with R134a, R600a and R600a/Al2O3—A Theoretical Comparative Study

Feiza Memet

Abstract In marine refrigeration there are currently used vapour compression refrigeration systems—technologies identified to be important energy consumers and traditional users of R134a refrigerant. It is important to tackle problems such as performance characteristics evaluation and adopting natural refrigerants instead of synthetics. Nano refrigerants are fluids with very good thermal properties, their use leading to the enhancement of vapour compression systems performance. In this paper it is analyzed the energy efficiency of such a system when a natural refrigerant (R600a) and a nano refrigerant (R600a/Al_2O_3) are alternatives to R134a. The thermodynamic analysis is established on the first and second laws of thermodynamics. The results show a performance improvement when using the nano refrigerant. Additional performance improvement it is reached when the refrigeration system is with sub cooling and super heating and when the volume concentration of Al_2O_3 nano particles increases.

Keywords Energy efficiency · Vapour compression refrigeration · Nano refrigerant

16.1 Introduction

Refrigeration, in particular vapour compression refrigeration systems (VCRSs), consume more energy and require more efforts to achieve energy efficiency, not only throughout replacing exiting refrigerant or components, but also by properly

---

F. Memet (✉)
General Sciences Department, Constanta Maritime University, Constanta, Romania
e-mail: feizamemet@yahoo.com

R. Tamas and C. A. Heiman (eds.), *Selected Papers from AToMS 2024*,
Lecture Notes on Multidisciplinary Industrial Engineering,
https://doi.org/10.1007/978-3-032-02452-7_16

dispersion of nano materials in the refrigerants—in order to obtain improved thermal properties of the refrigerant and boost the performance of the cycle [1, 2].

Traditional refrigerants show a poor thermal conductivity. By adding nanoparticles in conventional fluids will result nano refrigerants. Nano refrigerants can be obtained by dispersing nanoparticles (solid particles between 1–100 nm in size) into the refrigerant or by dispersing them into the compressor oil. In both cases, the final product is the nano refrigerant. Nano refrigerant enhances the heat transfer coefficient during evaporation and condensation, while the nano lubricant boosts tribology features which will lead to compressor performance improvement. Compared to conventional refrigerants, nano refrigerants show several benefits [3]: increased heat transfer surfaces between nano particles and the refrigerant, better dispersion stability, lower pumping power for the same amount of heat transfer, reduced particle clogging, versatile properties at different nano particles concentrations. The potential of nano refrigerants made them to play a key role in refrigeration sector [4]. The gains of nanotechnology in refrigeration are seen in improved refrigeration cycle performance, cooling capacity, thermal conductivity and lower power consumption [5].

Nano refrigerants can be prepared by two-step method or by one-step method [6]. Two-step method implies the production of dry nano sized powders using chemical or physical methods. During the one-step method, making and dispersing of nanoparticles are implemented simultaneously. The development of nano fluids and implicitly of nano refrigerants is based on the remark that the thermal conductivities of conventional refrigerants working in VCRSs possess very low values compared to metals or other fluids met in heat transfer processes; increasing the viscosity and thermal conductivity of the refrigerant responds to the strong requirement for new refrigerants showing excellent heat transfer characteristics [7].

Marine refrigerated transport is a sector which relies on VCRSs that are very often based on R134a refrigerant. This circumstance pushed this activity towards a more sustainable profile, being seen the important energy consumption and the high GWP (Global Warming Potential) of this refrigerant [8]. Despite its very good thermodynamic and thermo physical properties and null ODP (Ozone Depletion Potential), R134a is put on the list of greenhouse gases (GHGs) [9]. Different studies identified isobutane (R600a) to be an environmental friendly substitute for R134a [10, 11]. Their results revealed the decrease in compressor power consumption and increase in the Coefficient of Performance of the cycle. Flammability limitations in respect with the use of hydrocarbons (HCs), mostly in marine refrigeration, are not an obstacle if a strict risk assessment is performed and accurate maintenance procedures are involved. VCRSs designed for R134a require the replace of the old compressor with a HC one and the use of optimized capillary.

Even if international shipping is not included in the Paris Agreement, International Maritime Organization has its own strategy regarding lowering GHGs emissions from ships. For example, there is one general regulation regarding the deliberate escape of discharges of HFCs [12]. The trend in modern marine refrigeration is characterized by the use of natural refrigerants, such as R600a, an alternative to existing R134a. This paper deals with a theoretical thermodynamic analysis of a VCRS, based on the first and second laws of thermodynamics. It is analyzed the energy efficiency of this type of system when replacing R134a with R600a and R600a/Al_2O_3. By assessing the Coefficient of Performance and the exergy efficiency of the system will be possible to indicate a performance improvement when using the nano refrigerant R600a/Al_2O_3. Additional performance gain is reached when the refrigeration system is with sub cooling and super heating and when the volume concentration of Al_2O_3 nanoparticles increases.

16.2 Materials and Methods

The simple cycle VCRS– meaning without super heating and sub cooling, is illustrated in Fig. 16.1 [13].

Thermophysical properties of nano refrigerants (NRs) are in correlation with the ones of the pure refrigerant and nano particles (NPs). The enthalpy of the working fluid can be found by the help of NR density. Mollier charts of pure refrigerants are useful for finding enthalpy values at cycle points. The algorithm applied for the system's characteristics is given below [14–16]. Density of nano refrigerant (ρNR) is calculated as follows:

$$\rho_{NR} = \omega_{NP}\rho_{NR} + (1 - \omega_{NP})\rho_{PR} \tag{16.1}$$

where: ω_{NP}—mass fraction of nanoparticle, %,
ρ_{NR}—density of nanoparticle, kg m^{-3},
ρ_{PR}—density of pure refrigerant, kg m^{-3}.

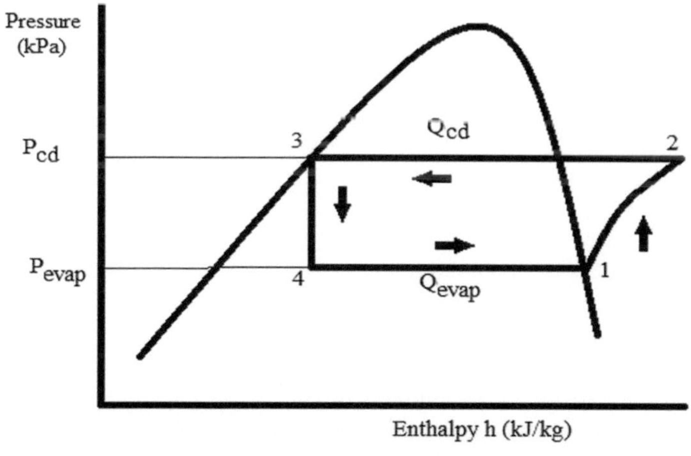

Fig. 16.1 P–h diagram of the refrigeration cycle: 1–2: isentropic compression in the compressor, 2–3: isobaric condensation in the condenser, 3–4: adiabatic expansion in the throttling valve, 4–1: isobaric evaporation in the evaporator.

Effective thermal conductivity of the nano refrigerant (K_{efNR}) can be determined as:

$$\frac{K_{efNR}}{K_{PR}} = \frac{K_{NP} + 2K_{PR} + 2\Phi_{NP}(K_{NP} - K_{PR})}{K_{NP} + 2K_{PR} - 2\Phi_{NP}(K_{NP} - K_{PR})} \tag{16.2}$$

where: K_{PR}—thermal conductivity of pure refrigerant, $Wm^{-1}K^{-1}$,
K_{NP}– thermal conductivity of nanoparticles,
Φ_{NP}– volume concentration (%).
While the volume concentration of nanoparticles (Φ_{NP}) is found with:

$$\Phi_{NP} = \frac{m_{NP}/\rho_{NR}}{\frac{m_{NP}}{\rho_{NR}} + \frac{m_{PR}}{\rho_{RR}}} \tag{16.3}$$

where:m_{NP}—mass of nanoparticle, g,
m_{PR}—mass of pure refrigerant, g.
Viscosity of nano refrigerant (μ_{NR}) is assessed with:

$$\mu_{NR} = \mu_{PR}(1 + 2.5\Phi_{NP}) \tag{16.4}$$

where:μ_{PR}—dynamic viscosity of pure refrigerant, Ns/m^2.
Analysis based on the first law of thermodynamics reveals the equations given below. Cooling capacity (q_{evap}) is calculated by:

$$q_{evap} = h_1 - h_4 \tag{16.5}$$

whereh—specific enthalpy, kJ/kg.
The specific power consumed by the compressor (wc) is given by:

$$w_c = h_2 - h_1 \tag{16.6}$$

Coefficient of Performance (COP) is calculated by:

$$COP = \frac{q_{evap}}{w_c} = (h_1 - h_4)/(h_2 - h_1) \tag{16.7}$$

The first law analysis is limited to assess only COP and states that energy cannot be created or destroyed. The second law deals with the quality of energy; it is useful in the evaluation of the quality of energy in actual vapour compression cycles because of irreversibilities encountered in the components of the systems (given by the fluid friction and heat transfer across a finite temperature difference). This is why energy and exergy analysis are used to improve thermal performance of VCRS. The second law introduces the concept of exergy (or the availability of a system). Exergy loss is a strong criterion when aiming system performance improvement. The improvement is gained if exergy losses are less, since the exergy efficiency depends on the available energy destruction. Mathematical relations specific to the second law of thermodynamics are provided in the following [17–20]. The exergy is expressed as:

$$ex = (h - h_0) - T_0(s - s_0) \tag{16.8}$$

where: s—entropy, kJ/kgK,
0—surrounding.

Exergy destructions in the components of the system should be calculated as follows. In the compressor:

$$Ex_{des.c} = Ex_1 - Ex_2 + \dot{W}_c \tag{16.9}$$

where: Ex—exergy rate, kW,
$\dot{W}_c$—power consumed by the compressor, kW.

In the condenser:

$$Ex_{des.cd} = Ex_2 - Ex_3 + \dot{Q}_c(1 - T_0/T_c) \tag{16.10}$$

where: $\dot{Q}_c$—condenser capacity, kW,
T_c–absolute condensation temperature, K.

In the throttling valve:

$$Ex_{des.th.v} = Ex_3 - Ex_4 \tag{16.11}$$

In the evaporator:

$$Ex_{des.evap} = (Ex_4 - Ex_1) + \left(1 - T_0/T_{evap}\right)\dot{Q}_{evap} \tag{16.12}$$

where: T_{evap}—absolute evaporation temperature, K,
$\dot{Q}_{evap}$—refrigeration capacity, kW.

The total exergy destruction in the cycle in determined as:

$$Ex_{des.TOT} = Ex_{des.c} + Ex_{des.cd} + Ex_{des.evap} + Ex_{des.th.v} \tag{16.13}$$

The exergy efficiency can be determined as:

$$\eta_{ex} = 1 - \left(Ex_{des.TOT}/\dot{W}_c\right) \tag{16.14}$$

16.3 Results and Discussion

The computation is developed on the assumed data: aluminium oxide nanoparticles of 10 nm nominal size; dead state: $t_0 = 30$ °C, $p_0 = 100$kPa; evaporation temperature: –10 °C to –20 °C; condensation temperature: 30–50 °C; mass fraction: 0.10%; volume concentration: 3–10%.

Figure 16.2 shows the variation of evaporation temperature (t_{evap}) on COP at constant condensation temperature ($t_c = 45$ °C), for a mass fraction of 0.10% Al2O3 nanoparticles.

It is observed that COP values increase together with the increase of the evaporation temperature, for all the considered refrigerants. The best values for this important efficiency criteria are found in the case of the nano refrigerant (R600a + Al2O3) and the lowest ones are registered for R134a.

It results that in the base of the first law analysis the replacement of R134a with a hydrocarbon (R600a) leads to a performance improvement. A better improvement is gained when replacing R134a with the hydrocarbon based nano refrigerant (R600a + Al2O3). COP increase is possible due the fact that higher evaporation temperatures lead to a gain in the refrigerating effect due to the increase of the refrigerant latent heat value and the reduction in the compressor work.

The comparison between exergy efficiency against evaporation temperature at constant condensation temperature is shown in Fig. 16.3 (NP mass fraction in still 0.10%). It is seen that a decrease of exergy efficiency is observed when the evaporation temperature increases. This trend is the same for the three refrigerants. In the considered range of evaporation temperature values, the exergy destruction ratio increases leading to an exergy efficiency decrease. At lower evaporation temperatures the nano refrigerant shows a greater efficiency.

Figures 16.4 and 16.5 are used to demonstrate the effect of condensation temperature (tc) on COP and exergy efficiency at constant evaporation temperature (t_{evap} = 0 °C) and 0.10% Al2O3 NP mass fraction. It is observed from Fig. 16.4 that for the three refrigerants COP decreases with the increase of condensation temperature. By keeping constant the evaporation temperature and increasing the condensation

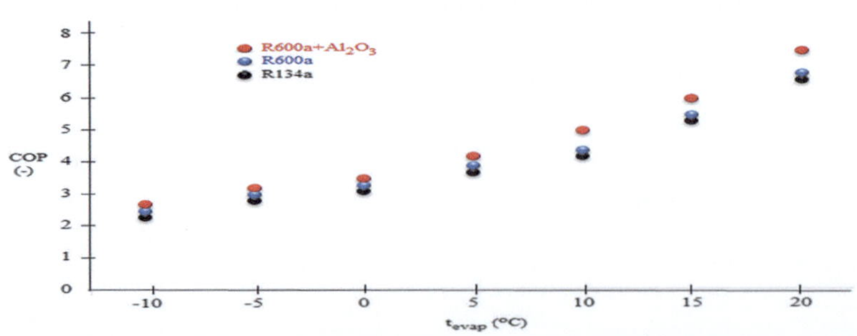

Fig. 16.2 COP versus evaporation temperature at 45 °C condensation temperature and 0.10% Al_2O_3 mass fraction

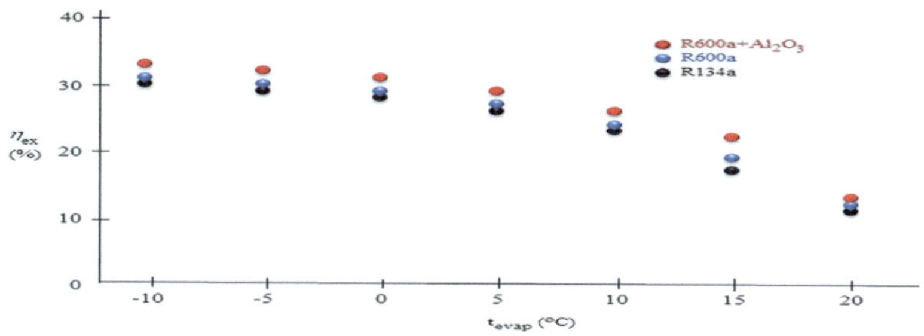

Fig. 16.3 Exergy efficiency versus evaporation temperature at 45 °C

temperature the cooling effect will decrease with a negative impact on COP. As shown in Fig. 16.5 exergy efficiency decreases with the increase of condensation temperature due to the fact that high condensation temperature value imply a greater exergy loss. In these two last situations it is also observed that R600a + Al2O3 working cycle shows the best performance, mostly for low condensation temperatures.

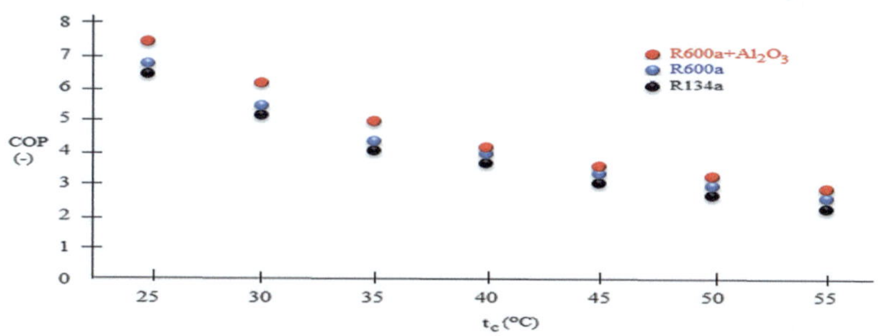

Fig. 16.4 Effect of condensation temperature on COP at 0 °C evaporation temperature and 0.10% Al_2O_3 NP mass fraction

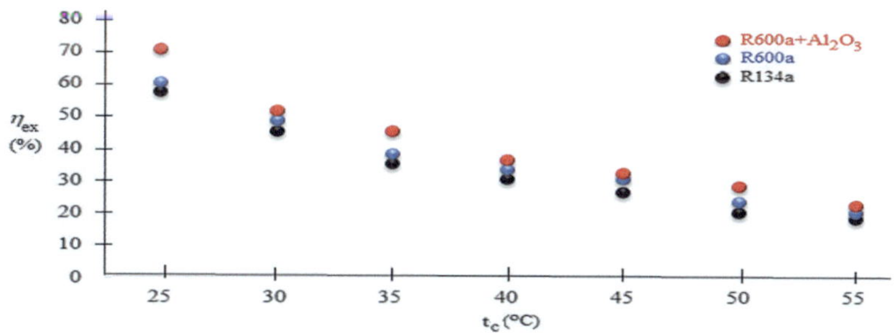

Fig. 16.5 Exergy efficiency influenced by the condensation temperature variation at 0 °C evaporation temperature and 0.10% Al_2O_3 NP mass fraction

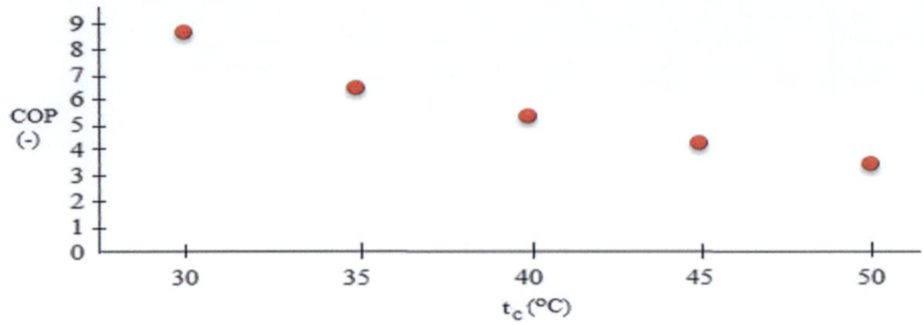

Fig. 16.6 Effect of condensation temperature on COP at evaporation temperature of 0 °C in the 5 °C super heating/sub cooling case

In the following the focus will be on the cycle working with nano refrigerant in order to investigate the performance of the system in different situations. The variation of COP with condensation temperature (t_c) at constant evaporation temperature ($t_{evap} = 0$ °C) and 5 °C super heating/sub cooling is provided in Fig. 16.6. COP results where found to be higher in the super heating/ sub cooling case (Fig. 16.6) compared to the no super heating/ sub cooling case (Fig. 16.4).

Figures 16.7 and 16.8 show the improvements in COP of the refrigeration system when the nano refrigerant is the working fluid and when comparing no super heating/ sub cooling and super heating/ sub cooling cases. Super heating and sub cooling introduced in the cycle will contribute to the increase of the refrigeration effect with good impact on COP values.

A further increment in COP with the increase of nanoparticle volume concentration is captured in Fig. 16.9. As the nanoparticle volume concentration increases COP is improved due to the lower amount of energy required by the compressor and to a gain in the refrigeration effect.

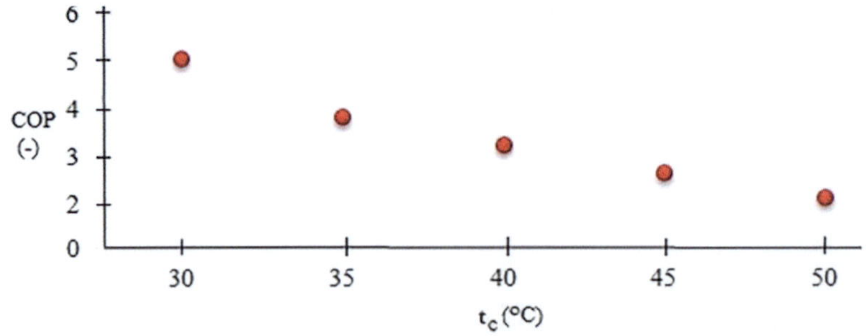

Fig. 16.7 Effect of condensation temperature on COP at evaporation temperature of 0 °C in the 5 °C super heating/sub cooling case

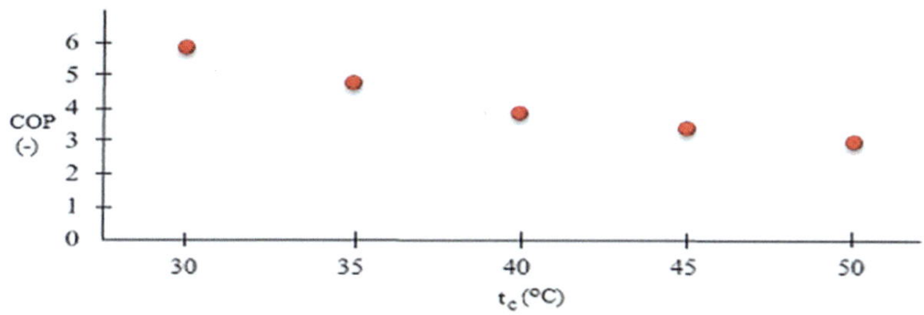

Fig. 16.8 Effect of condensation temperature on COP at −10°C evaporation temperature with 5 °C super heating/sub cooling

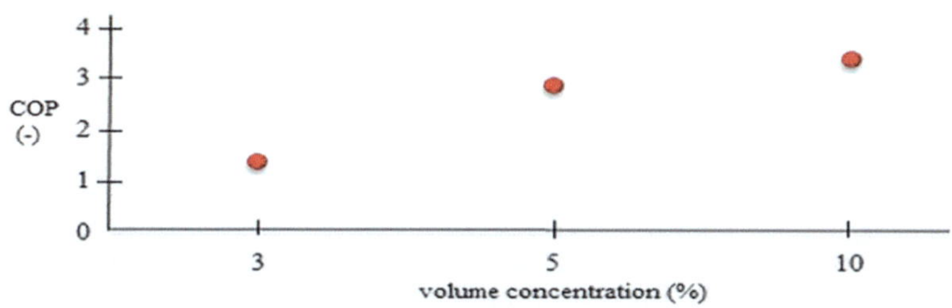

Fig. 16.9 COP behavior for different volume concentrations

16.4 Conclusions

In the lack of sub cooling and super heating processes, the highest COP value is of 7.4—obtained for the evaporation temperature of 20 °C and condensation temperature of 45 °C. COP increase is of 14% when using nano refrigerant and 3% when using the hydrocarbon. The best exergy efficiency is obtained also in the case of the nano refrigerant (70%) – when the evaporation temperature is 25 °C. The increases in this performance parameter are 21% (R600a/Al2O3) and 2% (R600a). When the evaporation temperature is 0 °C and the condensation temperature is 30 °C the highest COP value is of 8.6—obtained for the nano refrigerant use and when there is 5 °C sub cooling/ super heating degree. The COP gain brought by the sub cooling and super heating is of 41%. In the case of 5 °C sub cooling/ super heating COP values decrease with the decrease of the evaporation temperature. By decreasing the evaporation temperature from 0–10 °C, COP decreases with 32%. The performance of the system might be also enhanced by increasing the NP volume concentration: 148% COP increase when the NP volume concentration increases from 3 to 10%.

Acknowledgements Feiza Memet thanks to Constanta Maritime University for financing the publishing of this article from its research funds.

References

1. Mishra, R.S.: Vapour compression refrigeration systems using nano materials mixed with R718 in secondary circuit of evaporator for enhancing thermodinamic performances. Int. J. Res. Eng. Innov. **1**(3), 37–48 (2017)
2. Sakhir, A.S., Mahmoud, R.K.: The effect of nano-particles concentration Al_2O_3 on the performance in compression refrigeration system. Technium **3**(4), 67–80 (2021)
3. Soliman, A.M.A., Rahman, A.K.A., Ookawara, S.: Enhancement of vapour compression cycle performance using nanofluids. J. Therm. Anal. Calorim. **135**(2), 1507–1520 (2018)
4. Ajayi, O.O., Ukasoanya, D.E., Ogbonnaya, M., Salawu, E.Y., Okopujie, I.P., Akinlabi, S.A., Akinlabi, E.T., Owoeye, F.T.: Investigation of the effect of R134a/Al2O3 - nanofluid on the performance of a domestic vapour compression refrigeration system. Procedia Manuf. **35**, 112–117 (2019)
5. Babarinde, T.O., Akinlabi, S.A., Madyira, D,M.: Enhancing the performance of vapour compression refrigeration system using nano refrigerants: a review. IOP Conf. Ser.: Mater. Sci. Eng. **413** 012068, 1–8 (2018)
6. Vandaarkuzhali, S., Elansezhian, R.: Performance evaluation of air conditioning system using nanofluids. Aust. J. Basic Appl. Sci. **9**(7), 1–10 (2015)
7. Ogbonnaya, M., Ajanyi, O.O., Waheed, M.A., Oyedepo, S.O.: Review on the energy and exergy analysis of vapour compression refrigeration system using nanolubricant. J. Phys.: Conf. Ser. **1378**(4) 042067, 1–17 (2019)
8. Minetto, S., Fabris, F., Marinetti, S., Rossetti, A.: A review on present and forthcoming opportunities with natural working fluids in transport refrigeration. Int. J. Refrig. Int. J. Refrig. **152**, 343–355 (2023)
9. Hmood, K.S., Apostol, V., Pop, H., Badescu, V., Pop, E.: Drop-in and retrofit refrigerants as replacement possibilities of R134a in domestic/commercial refrigeration and automobile air conditioner applications. J. Ther. Eng. **7**(7), 1815–1835 (2021)
10. Akyol, S., Das, M., Alatas, B.: Modeling the energy consumption of R600a gas in a refrigeration system with new explainable artificial intelligence methods based on hybrid optimization. Biommetrics **8**(5), 1–30 (2023)
11. Hastak, S., Kshirsagar, J.M.: Comparative performance analysis of R600a and R436a as an alternative of R134a refrigerant in a domestic refrigeration. IOP Conf. Ser.: Mater. Sci. Eng. **377**, 012047, 1–9 (2018)
12. Glavan, I., Poljak, Mustac, I.M., Lonic, I.: Natural refrigerants on board Marine Vessels. Pomorski Zbornik **62**(1), 43–56 (2022)
13. Suresh, V., Baskaran, A., Manikandan Subaramanian, K.: A performance study of vapour compression refrigeration system using ZrO2 nanoparticle with R134a and R152a. Int. J. Sci. Res. Publ. **6**(12), 410–421 (2016)
14. Dilawar, M., Qayoum, A.: Simulation of vapour compression air conditioning system using Al2O3 based nanofluid refrigerant. J. Ther. Eng. **9**(5), 1307–1323 (2023)
15. Odunfa, M.K., Oseni, O.D.: Numerical simulation and Performance assessment of a nanoparticle enhanced vapour compression refrigeration system. J. Energy Power Eng. **9**(11), 33–49 (2021)
16. Hamad, A.J.: Experimental investigation of vapor compression refrigeration system performance using nano-refrigerant. Wasit J. Eng. Sci. **2**(2), 12–27 (2014)
17. Babarinde, T.O.: Exergy analysis of refrigeration system using R600a with TiO_2 nano lubricant. In: Proceedings of the International Conference on Industrial Engineering and Operation Management, Washington DC (2018)
18. El Sayed, R., El Morsi, M., Mahmoud, N.A.: Thermodynamic analysis of a simple refrigeration cycle using hydrocarbon refrigerants as substitute to R22. Int. J. Adv. Eng. Manag. Res. **2**(2), 245–274 (2017)
19. Ahamed, J.U., Saidur, R., Masjuki, H.H.: A review on energy analysis of vapor compression refrigeration system. Renew. Sust. Energ. Rev. **15**(3), 1593–1600 (2011)

20. Karacayli, I., Simsek, E.: Exergetic investigation of the effects of superheating and subcooling on performance of a vapor compression refrigeration cycle. Eur. Mech. Sci. **4**(4), 152–157 (2020)

Chapter 17
Marine Auxiliary Engine Piston Assembly Solidworks Study–Dynamic Behaviour Analysis

Cătălin Faităr, Ionuţ Voicu, Fănel-Viorel Panaitescu, Andrei Scupi, Mariana Panaitescu, Nicolae Buzbuchi, and Anastasia Duşe

Abstract Marine auxiliary engines have specific funcional characteristics that make them very different from other types of main engines commonly used in the marine industry. It is well known that most of them are four-stroke engines with semi-fast or fast rotation speeds. Understanding their operating principles is essential for designing the optimal naval energetic system configuration. This study aims to carry out a detailed analysis of a certain type of marine auxiliary engine, which is widely used in the maritime industry, in various configurations for several decades now.

Keywords Four-stroke · Piston · Simulation · SolidWorks

17.1 Introduction

Some 105 years after the entry of the Selandia into service, considered the first seagoing vessel to be powered by engines, diesel engines almost completely dominate the ship propulsion market. The essential sectors of maritime transport have long abandoned steam turbines and generators, being replaced by low and medium speed engines, which are widely used on board container ships, bulk carriers, VLCCs and passenger ships. Even the last bastion of steam (turbines) in ship propulsion–LNG tankers–has been conquered by dual fuel diesel engines that have the ability to run on a mixture of diesel fuel and gas resulting from the evaporation of the liquefied natural gas it carries [1, 2].

C. Faităr (✉) · I. Voicu · F.-V. Panaitescu · A. Scupi · M. Panaitescu · N. Buzbuchi
Department of Mechanical and Environnmental Engineering Science, Maritime University of Constanţa, Constanţa, Romania
e-mail: catalinfaitar@yahoo.com

A. Duşe
Departament of Navigation, Constanta Maritime University, Constanta, Romania

R. Tamas and C. A. Heiman (eds.), *Selected Papers from AToMS 2024*,
Lecture Notes on Multidisciplinary Industrial Engineering,
https://doi.org/10.1007/978-3-032-02452-7_17

To highlight the progress made in the 75 years between the construction of these two ships whose propulsion was based on four-stroke engines, it must be specified that the engines on board the Queen Elizabeth had a bore of 580 mm, a stroke of 640 mm, and each cylinder individually develops a maximum continuous power of 1213 kW, more than the entire engine on board the ship Selandia. The first ship equipped with propulsion engines proved its efficiency by being able to sail for 30 years. The ship's cruising speed was kept constant at 10.5 knots, the propulsion engines proving their efficiency all this time. In 1942, the ship was sunk off the coast of Japan under unknown conditions.

In the period of 1950–1951 the Danish MAN engine building company began to test and introduce high-pressure supercharging for medium-speed four-stroke engines for which the required and the obtained charge pressure was about 2.3 bar. Progressive advances and the efficiency of these supercharging systems allowed this principle to be applied to two-stroke engines in the mid-1950s. The contributions brought by the research in this field allowed the turboblowers to reach yields higher than 70%.

Along with the concept of the supercharged engine, the market for building turbochargers (tubines) for marine engines also developed, at first only for four-strokes. The company that researched and progressed in this regard was ABB Turbo-Systems, through its subsidiary Brown Boveri (BBC), originating in Switzerland. Thus, the BBC launched the first series of turbofans in 1944, called VTR.0. This marked the end of an era where each turbine was custom built for an individual engine or installation.

Thus, it was possible to build and install turbines that could be used in a variety of applications, being able to serve engines that developed powers in the range of 370–14,700 kW. Until then, turbochargers were custom-built only for four-stroke engines, but from 1944 it was possible to use them in two-stroke engines that were also equipped with scavenging air pumps.

Another important step in the progress of four-stroke marine engines was recorded when the use of heavy fuel was switched to. Currently, almost all marine engines, whether they are slow two-stroke or medium-speed four-stroke, also work with solid fuels with up to 700 cSt at 50 0C, work is still being done on the possibility of using these types of fuels and in the case of fast four-stroke engines. The main barrier to the use of these types of fuels was the fact that they deteriorate during bunkering, but technological advances have allowed adaptation to today's market requirements.

Although the marine diesel propulsion market is dominated by slow two-stroke engines with direct coupling and cross capr system or medium or high-speed four-stroke engines equipped with a reversing reduction system, there is a hybrid variant that has always been used by the market in Japan and the East Asian area, especially by coastal operators or those in the fishing industry. It's about slow four-stroke engines.

17.2 Materials and Methods

In the period of 1950–1951 the Danish MAN engine building company began to test and introduce high-pressure supercharging for medium-speed four-stroke engines for which the required and the obtained charge pressure was about 2.3 bar. Progressive advances and the efficiency of these supercharging systems allowed this principle to be applied to two-stroke engines in the mid-1950s. The contributions brought by the research in this field allowed the turboblowers to reach yields higher than 70% [3, 4].

Engines from the Wärtsilä 32 range (benchmark bore value) have been the preferred choice of shipyards, operators and owners since the 1980s, with over 5100 engines delivered to the marine market alone. The Wärtsilä 32 is available in various construction variants with configurations from 6 to 16 cylinders and a developed power between 3 and 9.3 MW at rated speed values between 720 and 750 RPM. It has best-in-class power efficiency and fuel economy, and can be used in a wide operating range. With proven reliability and low consumption of consumables, the Wärtsilä 32 represents the most efficient solution throughout the ship's life cycle, presenting the following advantages [4]:

- Long-term use in the maritime field;
- High reliability;
- High power density;
- Reduced fuel consumption on a wide range of loads;
- Runs on HFO, MDO and liquid biofuels (including LNG).

The Wärtsilä 32 engine has a proven track record in a wide range of marine applications. It is used for specific applications as prime movers in both electric and diesel powertrain configurations, but is also widely used as an auxiliary drive. It can be optimized for constant speed or along a combinatorial load curve. In the commercial fleet, typical applications include use as the main engine on various types of tankers and container ships, as well as in various configurations for passenger ships, especially in electric propulsion systems. In the offshore sector, the reliability of the Wärtsilä 32 has made it the most popular medium-speed engine for both OSVs (offshore support vessels) and drillships. Similarly, in the cruise ship and ferry sector, the Wärtsilä 32 has proven to be the most favored engine in its size class. In ship power systems Wärtsilä 32 is widely used in all ship categories where a high power load is required [5, 6] (Fig. 17.1).

The excellent flexibility of the fueling mode allows the Wärtsilä 32 to operate on HFO, MDO and liquid biofuel at a wide range of fuel viscosities from 2.0 cSt to 730 cSt for HFO, for example (at 50° C / 122 °F). The engine is able to run efficiently and economically on low sulfur fuels (<0.1% S), making it suitable for operation in Controlled Emission Areas (SECA). The engine can also be equipped with an SCR catalyst, such as Wärtsilä NOR (nitrogen oxide reducer), which can reduce NOX emissions by up to 95%. This means that the engines are already IMO Tier III compliant. The Wärtsilä 32 naturally meets IMO Tier II regulations [7, 8].

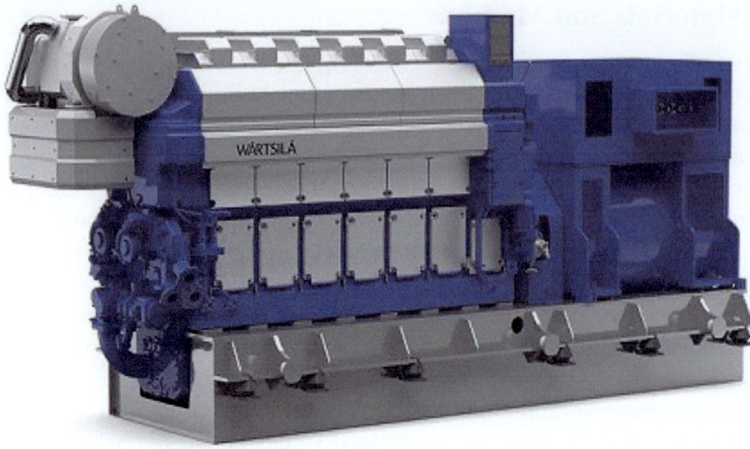

Fig. 17.1 Typical view of an Wärtsilä 32 auxiliary engine [6]

The Wärtsilä 32 engine is equipped with a variable intake valve shut-off unit (abbreviated VIC). This makes it possible to apply early intake valve closure at high load, which in turn allows for reduced NOX levels and reduced fuel consumption. By switching to late intake valve closing, good part load and transient performance is ensured. Overall operational benefits include improved performance at part loads, reduced smoke levels and improved load regulation. The engine control system incorporates automatic monitoring and control for optimal operating efficiency.

The main technical characteristics of the engine presented above are the following [9]:

- Bore: 320 mm;
- Stroke: 400 mm;
- Piston displacement (cylinders): 32.2 L per cylinder;
- Number of valves: 2 exhaust valves, 2 intake valves per cylinder head;
- Speed: 720–750 rpm;
- Average piston speed: 9.6–10 m/s;
- Dimensional characteristics:
- Power (as main engine): 7223 HP;
- Power (as auxiliary engine): 5040 kW (6050 kVA).

It should be remembered that the piston is a mobile organ that ensures the evolution of the engine fluid necessary for the production of mechanical work, and the part that contributes to its movement (segments and bolt in engines in the case of engines in 4τ, especially segments, piston rod and cross head for the engines in 2τ) they compose, together with the piston, what is called the piston group, ensuring the assembly and operation of the respective organ.

Fig. 17.2 Stepts to achieve the Isometricmodel of the Piston assembly of the Wärtsilä 32 engine (Solidworks Softwear 2025)

17.3 Results and Discussion

17.3.1 Study of the Piston Assembly

In the present case, the study of the piston assembly consists in making the actual model and defining the geometric parameters of the components to be simulated and analyzed. The designed components will be the piston, connecting rod and piston pin. In the first phase, these components are drawn individually, and the assembly is defined geometrically using the specific commands in Solidworks. A special emphasis is placed on the connections in the component elements. It is vital that the result of this action is the creation of a model as close as possible to the real assembly (Fig. 17.2).

Following the adhesion/strength test of the chromium layer we can conclude that the sample has satisfactory characteristics and can be classified as HF4 according to the method used.

17.3.2 Bodies and Materials

The next step is to establish the material properties of the components designed above. In this regard, it is established that the piston and connecting rod are made of cast aluminum alloy, and the piston pin is made of an AISI 3140 steel alloy (Fig. 17.3).

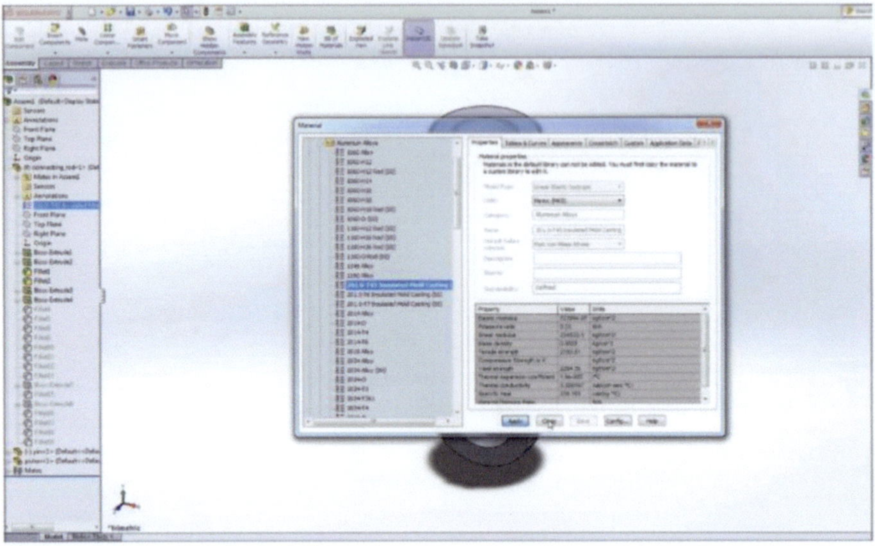

Fig. 17.3 Steps to achieve the isometric model of the piston assembly of the Wärtsilä 32 engine (Solidworks Software 2025)

17.3.3 Interactions

In this stage of the simulation the ways in which the surfaces of the component elements within the assembly interact (or come into contact) are established. In establishing these conditions, the reference element is the piston connecting bolt, being the component that connects the other two component elements, therefore it is very important to establish the surfaces with which they come into contact. To create the most realistic image from the control tree, the bolt, its internal and external faces/ surfaces are chosen and the contact ratio with the other components is established. This procedure is repeated for the other piston surfaces (two more times) (Fig. 17.4).

17.3.4 Mesh and Run

The next stage includes the discretization (meshing) stage and, depending on the obtained model data, the creation of the necessary framework for running the actual simulation of how various factors act on the piston assembly. An important step is to set the forces acting on the components within the assembly. One of the "external" forces acting on the assembly is set to be the one induced by the engine's crankshaft. Another force set as having a direct action on the assembly is the pressure force acting on the piston head (Fig. 17.5).

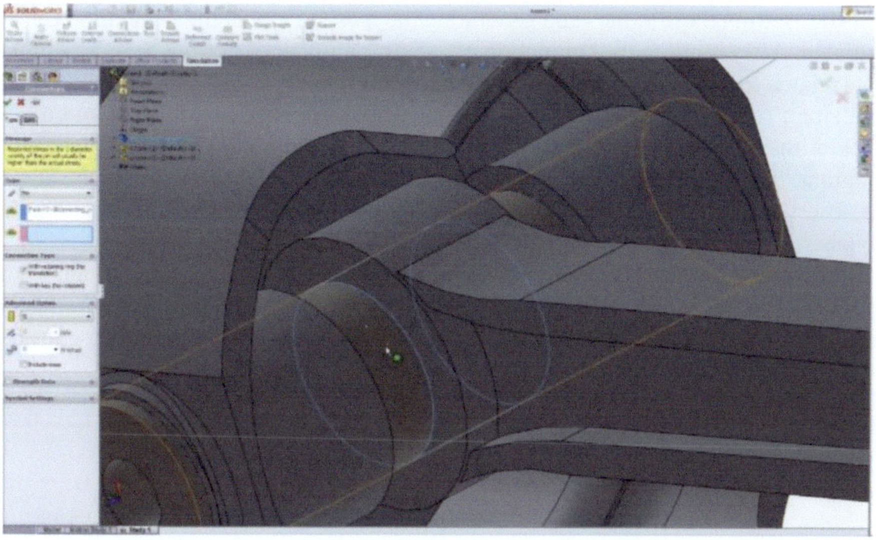

Fig. 17.4 Connections set for the piston assembly (Solidworks Software 2025)

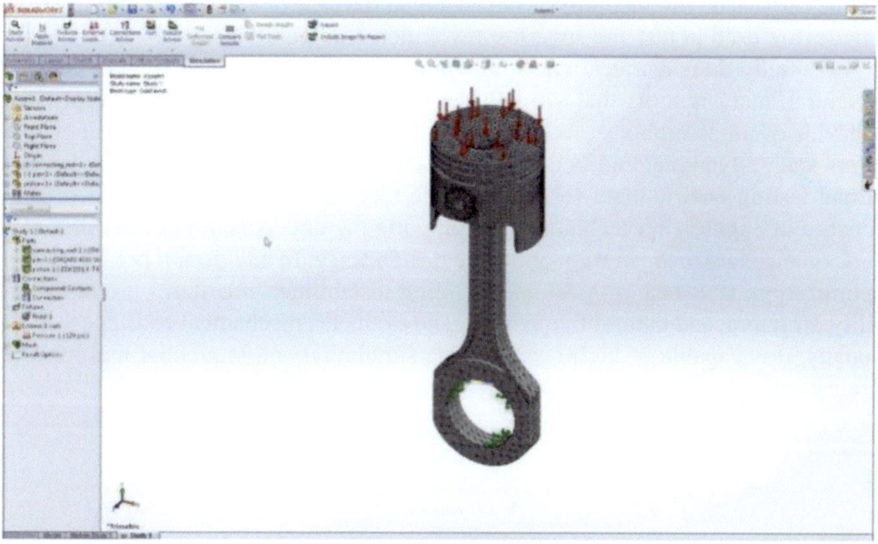

Fig. 17.5 Meshing of the piston assembly with the setting of the external loads (Solidworks Software 2025)

After setting these values and generating the meshed model the simulation can be initiated. After running the simulation, there are a number of options to represent the data related to the simulation, the default ones being stress (von Mises), displacement and strain (equivalent). Higher stress concentrations on the piston head and lower neck of the connecting rod can be seen on the stress diagram. On the displacement diagram, the maximum value of the displacement recorded on the piston head can be observed, which is equal to 0.4 mm. On the strain diagram it can be seen that the connecting rod has the highest specific values (Fig. 17.6).

Solidworks offers additional analysis capabilities. One of the most important being those related to the safety factor. This safety factor gives a detailed picture of the most exposed areas of the assembly, that is, those areas that can fail under extreme stresses (Fig. 17.7).

17.4 Conclusions

The basic side of the four-stroke engines used in the maritime field remains their use in generating sets, the only difference being that in the case of large ships four-stroke engines are used with medium speeds, while in the case of smaller ships four-stroke engines are used in fast and superfast revolution configurations.

This study demonstrates that Solidworks Simulation is a user-friendly suite of structural analysis tools that virtually tests CAD models to forecast the physical behavior of an assembly in real life using Finite Element Analysis. The portfolio offers static, dynamic, and non-linear analysis capabilities. This is a user-friendly virtual testing environment for linear, static motion that is time-based. It offers a concurrent engineering technique that supports the understanding of the product's performance and lifespan throughout the design stage. In addition, it permits design optimization, tests heat transfer and buckling instabilities, ascertains product durability, structure, and natural frequencies, and evaluates mechanical resistance. Additionally, it is capable of high-cycle fatigue simulation and sequential multiphysics simulations.

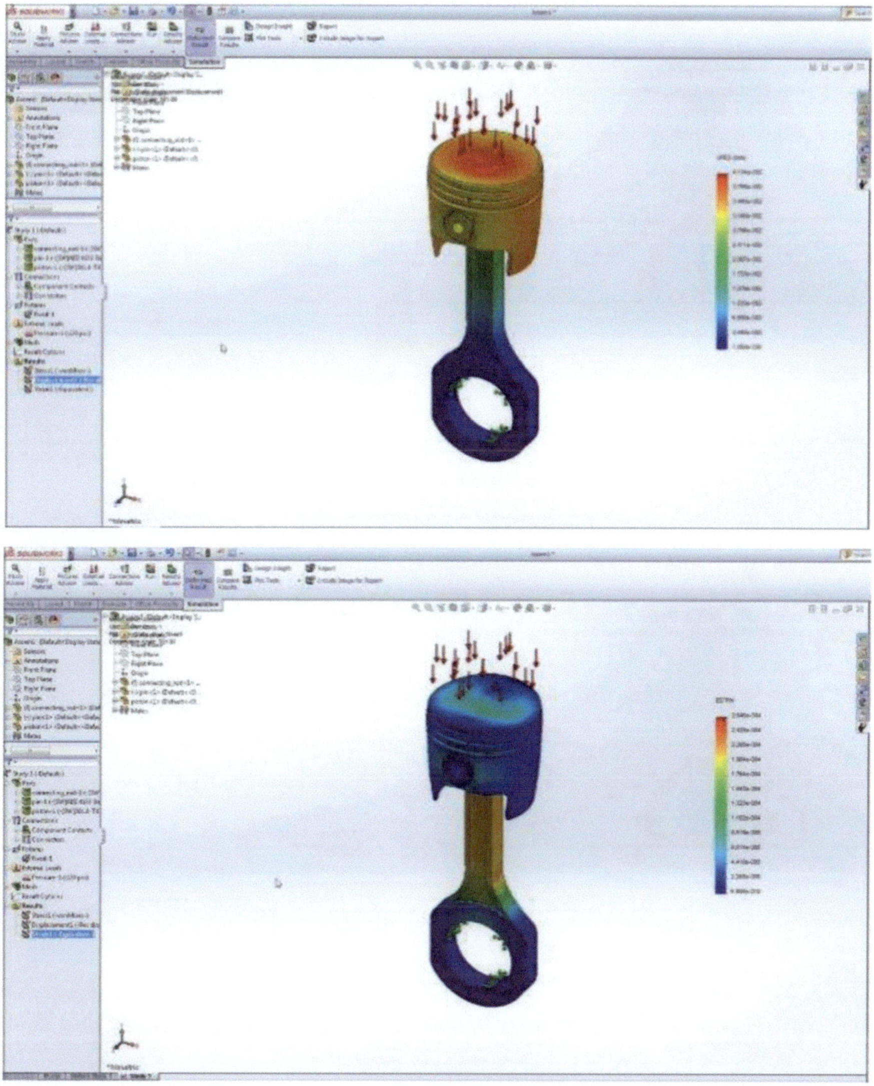

Fig. 17.6 Results for the deformations generated by the external loads (Solidworks Software 2025)

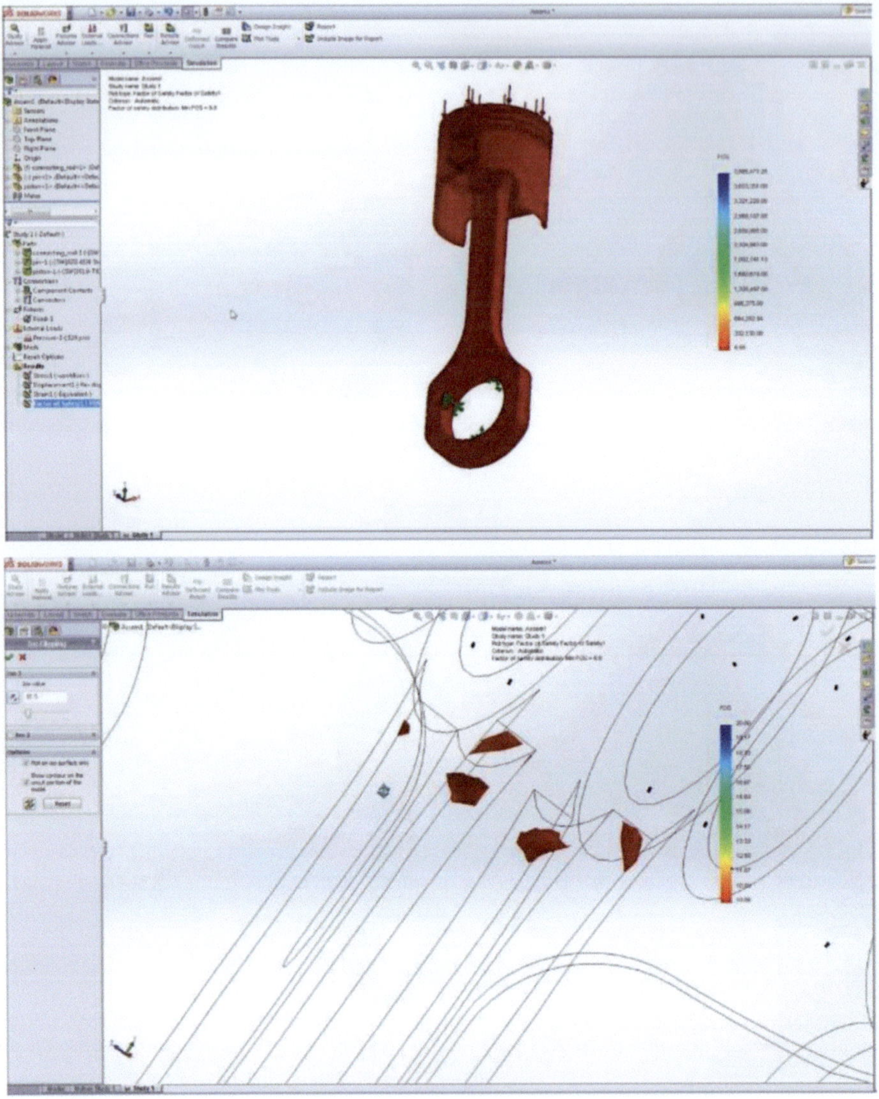

Fig. 17.7 Piston assembly safety factor analysis (Solidworks Software 2025)

References

1. Faităr C., Novac, I.: A new approach on the upgrade of energetic system based on green energy. A complex comparative analysis of the EEDI and EEOI. In: Proceedings of the ModTech International Conference Modern Technologies in Industrial Engineering, Iaşi, România, pp. 15–18 (2016)

2. Faităr, C., Rusu, E., Buzbuchi, N., Dumitrache, C.L., Stan, L.C.: The importance of measuring performance of the main naval engine and practical measurements methods, Mechanical Testing and Dignosis, vol. 12, no. 2, pp. 5–10, Dunărea de Jos University, Galați, România (2022)
3. Grimmelius, G.T.: Simulation auxillary models in marine engineering: from training to concept exploration, Compit, Hamburg, pp. 50–61 (2023)
4. Babicz, J.: Wartsila Encyclopedia of ship tehnology, Second Edition, Wartsila Corporation, pp. 105–121 (2015)
5. Skippon, K.: Piston Assembly Stress Analysis—Solidworks Tutorial, Unpublished (2014)
6. Cattarinussi, M., Leppaen, A., Konno, J., Frondelius, T.: Cylinder head design of experiments by using the Wartsila Digital Platform. J. Struct. Mech **50**(3), 330–332 (2017)
7. Vasilescu, M.V.: Advantages and disadvantages of different types of modern marine propulsions. J. Mar. Technol. Environ **2**, 57–63 (2018)
8. Hansen, S.V.: Performance monitoring of ships, technical university of Denmark, Department of mechanical engineering, pp. 82–88 (2011)
9. Wartsila.: Wartsila 32. Product Guide, Wartsila Engines, Warstila OY, Vassa, Finland, pp. 32–51 (2019)

Chapter 18
Experimental Set-Up and Prelimina+y Study on the Evolution of Tire-Road Contact Patch

Ioan-Cozmin Manolache-Rusu, Cornel Suciu, Ionuţ-Cristian Românu, and Andrei Cosmin Dumistrăcel

Abstract The present paper aims to propose an experimental set-up for determining the contact patch geometry and its area for a given tread design when changing more important parameters such as inflation pressure and load. The present paper describes the design, finite element method (FEM) analysis and construction of an experimental test rig that allows the for a quarter car model allowing the analysis of the contact patch displacement in static regime by optical image analysis methods. Such a study can contribute to the identification of best practices for increasing tire durability, fuel efficiency, pollution reduction, stability, handling and safety.

Keywords Contact patch · Experimental test-rig · Tire · Optical

18.1 Introduction

In order to transform the rotational motion of the wheel into the translational motion of the frame or body, the direct link between the majority of today's road vehicles and the road is via the tire. For this purpose, all reaction forces at the tire to road contact interface will be transferred to the axle guidance system and transmitted to the elastically suspended mass. The size and geometry of the tire-road contact patch influences the stress distribution at the interface, which has been highlighted by other authors in [1, 2], and thus directly influences the dynamic behavior, handling and performance of the vehicle. An inadequate contact patch will lead to high stresses that amplify the phenomenon of excessive and uneven tire wear, reduce stability through the occurrence of skidding, favoring an increase in accident rates.

A detailed study of the tire-to-road contact patch evolution under both static and dynamic loading conditions can contribute to increased tire durability, fuel efficiency,

---

I.-C. Manolache-Rusu · C. Suciu (✉) · I.-C. Românu · A. C. Dumistrăcel
Department of Mechanics and Technologies, Stefan cel Mare University, Suceava, Romania
e-mail: suciu@usm.ro

R. Tamas and C. A. Heiman (eds.), *Selected Papers from AToMS 2024*,
Lecture Notes on Multidisciplinary Industrial Engineering,
https://doi.org/10.1007/978-3-032-02452-7_18

reduced pollution, increased stability, vehicle handling and road traffic safety. Studies of tire-road contact have been previously carried out using different simplified models such as quarter, half or full car [3, 4]. They aimed at highlighting the vehicle behavior on different categories of roads with different grip coefficients depending on the type of road surface [5–7] and environmental conditions.

The analysis proposed in this paper can contribute to the development of new tire technologies and materials with optimized characteristics that provide increased grip and performance on a variety of road surfaces. The study itself aims, in a first phase, to develop a static test rig to determine the contact patch geometry and effective area sizes for different tire pressure values, as well as the contact patch displacement corresponding to different torque applied to the planetary shaft.

18.2 Experimental Set-Up

In order to achieve the proposed goal, in accordance with the aspects presented above, this sub-chapter presents the design and construction of an experimental test platform that can be used to analyze the evolution of the tire-road contact for a quarter car model.

Figure 18.1 shows the mechanical structure of the proposed experimental set-up, which has been designed in order to allow the analysis of the evolution of the contact patch by means of an optical method of analysis and subsequent graphical image processing.

The experimental test rig base structure consists of frame 3 and sub-frame 4, which are joined by means of a planar coupling in the form of a vertical axial guide. Frame 3 features a thick glass plate 1 at the top, on which a self-adhesive foil with a millimeter grid is glued. At the lower part of the frame there is a crossbeam on which the hydraulic jack 7 is positioned to allow for loading the tire-glass surface contact. Also, a load cell 8 is positioned between the frame and the hydraulic jack to allow monitoring of the applied load magnitude.

On the half-frame 4 there are positioned the elements of an independent suspension with a MacPherson-type swinging arm suspension, i.e. the elements of the running gear of a front axle vehicle with drive and steered wheels. Thus the lower arm 11 is positioned relative to the half-frame 4 by means of rubber-metal springs, and relative to the leaf carrier 5 by means of a ball joint. At the upper part, the shock absorber 6 and the suspension spring coaxially mounted and also resting on the half-frame 4 are shown. The experimental stand has also been fitted with the planetary shaft 10 which on the wheel side has a Rzeppa type homokinetic angular-axial joint and at the opposite end a recess for mounting a ½ inch square dynamometer to monitor the torque applied to the wheel.

The proposed positioning of the system differs from the classical position, being rotated by an angle of 180° around the Y-direction of the planetary shaft, in order to

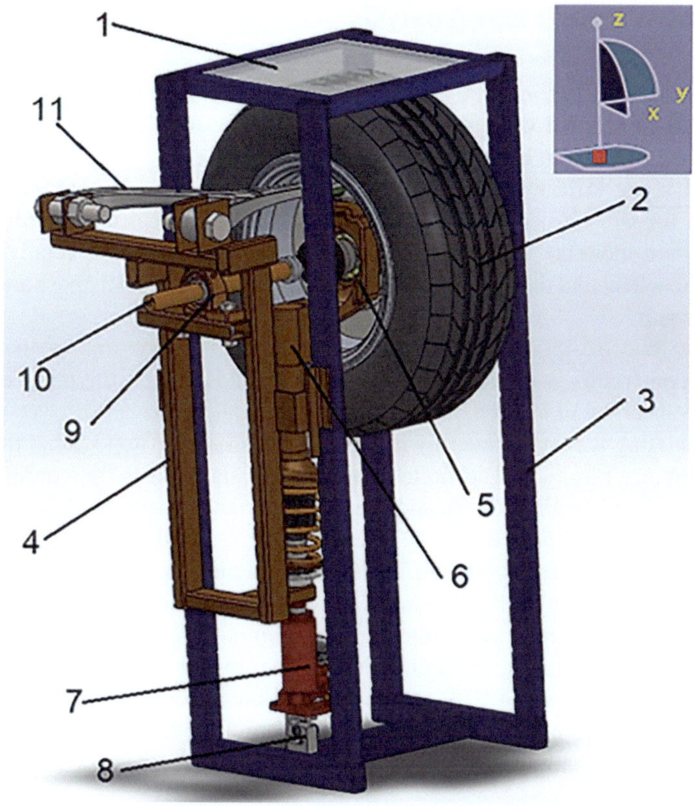

Fig. 18.1 Structure of the proposed test rig: 1–glass surface; 2–test wheel; 3–frame; 4–subframe-support; 5–steering knuckle; 6–shock absorber; 7–loading device; 8–load cell; 9–roll bearing; 10–planetary gear set; 11–control arm

analyze the contact patch more easily. Therefore, the proposed experimental test set-up allows to simulate the contact patch displacement in static mode and to evaluate the contact area by optical image analysis methods.

The way the test rig works involves the application of hydraulic pressure to the jack 7, the load being evaluated by the load cell 8. In this way, by means of the suspension elements 6, the tire 2 will fold on the support glass 1, leading to the appearance of a contact patch whose position, shape and size will be influenced by the loading conditions and the geometry of the components. By using real components in the construction of the stand, which are often used in today's vehicles, it was possible to simulate conditions very close to the real loading conditions.

18.3 Sub-Frame FEM Analysis

The modeling was carried out in SolidWorks environment, considering as material an AISI 1035 steel with ultimate tensile strength of 282.68 MPa and material characteristics shown in Fig. 18.2.

To verify the stiffness of the test equipment, a finite element analysis was performed for sub-frame 4 of the structure. This element was chosen for the analysis because it models the suspended mass (vehicle body), and the relative position between the wheel and the roadway is directly influenced by the stiffness and strength of this element.

A static load of 4500N was applied to the model on the end of the arm and clamping constraints were applied to the cylindrical bores on the other end of the subframe, as shown in Fig. 18.3.

The modeling was optimized by modifying the meshing step so that the convergence errors are minimal. The finite element analysis yielded the vonMises stress distribution shown in Fig. 18. 4a and the displacements along the direction of load application plotted in Fig. 18. 4b. By analyzing the obtained results, it was observed that the maximum stresses, of 256.7 MPa, are reached in the mounting region of the wheel support arm and sub-frame. The maximum value of vonMises stresses obtained from modeling is below the allowable limit of the considered material.

The displacements along load direction, shown in Fig. 18. 4b, illustrate that the maximum displacement, of 5.99 mm is reached on the end of the arm and do not influence the relative position between the wheel and the road.

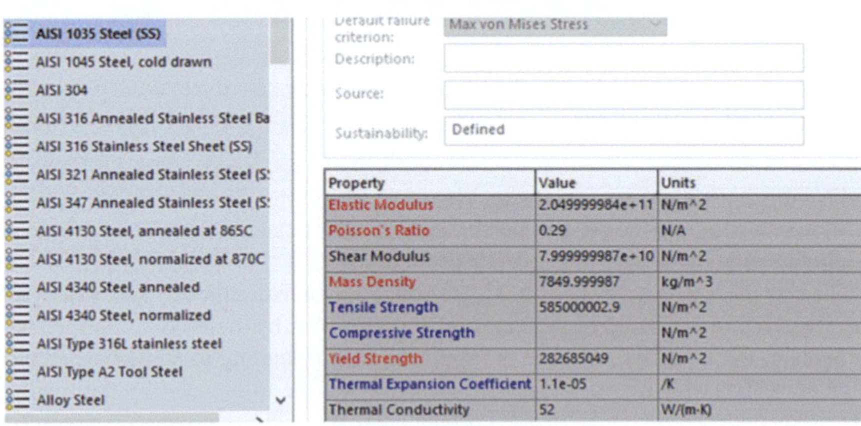

Fig. 18.2 Material parameters

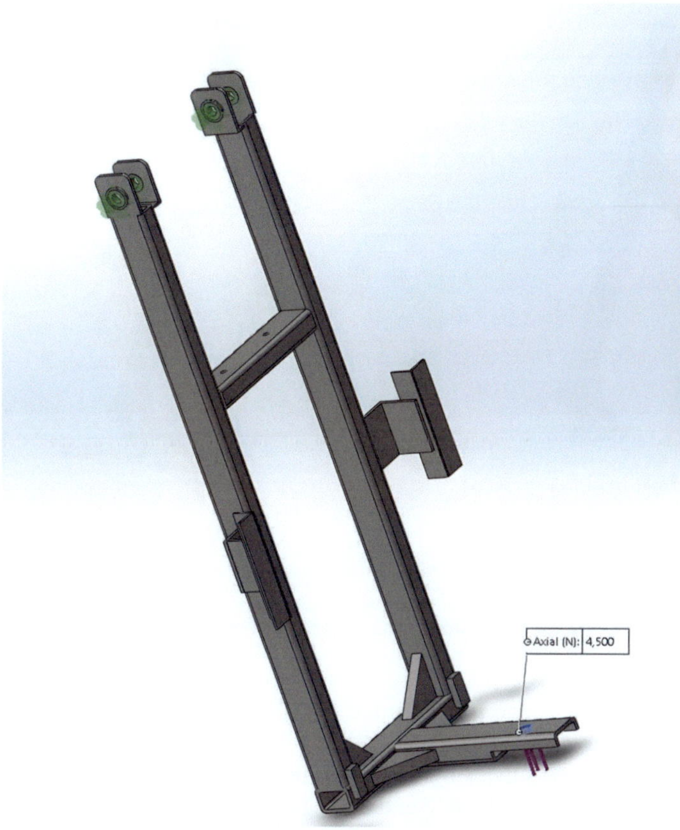

Fig. 18.3 Load and constraints applications on the 3D model

18.4 Experimental Methodology

In order to achieve the purpose of the proposed work, the built stand, illustrated in Fig. 18.5, uses a series of measuring instruments such as:

- A load cell fitted with a module displaying the load in kgf that allows the evaluation of normal load of the wheel. In a first stage, the load cell was calibrated by weighing three bodies of known mass. For each load level, the value indicated by the strain gauge was recorded, thus obtaining the dependence between applied load and stress. The experimental values allowed determining this dependence and subsequently the data were fed into the load cell module for load display.
- An optical transducer is attached to the frame of the test stand and is used for linear displacements used when measuring the relative distance between the raceway and the planetary shaft axis, which actually reflects the tire deformability (change in static wheel radius);

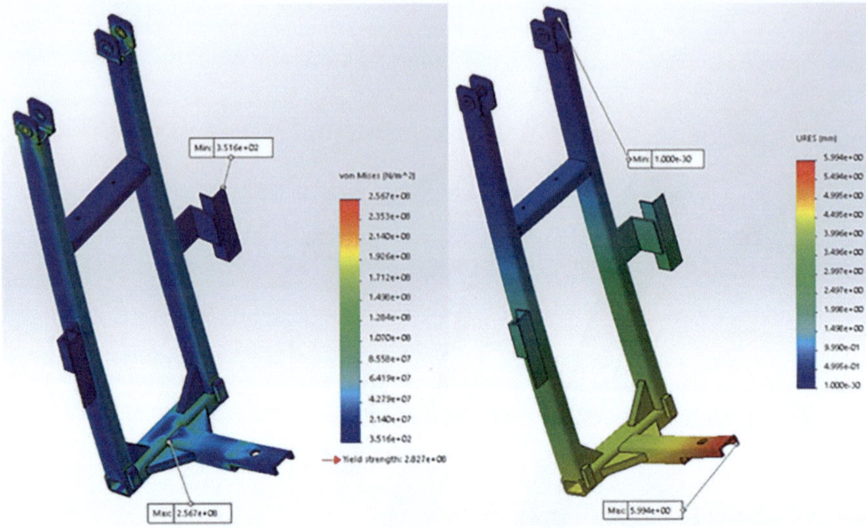

(a) (b)

Fig. 18.4 FEA results **a** vonMises stress state, **b** Displacements along load direction

- A digital pressure gauge to determine the tire inflation pressure, set for all test cases at 2.35 bar;
- A digital torque transducer for determining the torque applied to the planetary shaft. It is positioned on the free end of the planetary shaft and allows a controlled torque to be applied by attaching bodies of different masses;
- Imaging equipment to determine the shape and size of the contact patch between the wheel and the road.

18.5 Experimental Results and Analysis

For the proposed preliminary experimental study, normal loads between 50 and 450kgf were applied to the tire-flat surface contact. For each load value, the size and shape of the contact patch area, as well as normal approach between contact bodies were measured and recorded.

The results obtained for the shape and size of the contact patch between the tire and the road surface are shown in Fig. 18.6 a, b and c.

An image analysis was conducted on the obtained contact patches using ImageJ software. This allowed to determine the contact area boundary and surface area. Such an analysis is illustrated in Fig. 18.7, corresponding to a 50kgf normal load.

Fig. 18.5 Experimental set-up overview

For each load level, the contact area magnitudes were measured, and the following results were obtained: 1620.1 mm^2 for 50kgf normal load, 4899.3mm^2 for 200kgf normal load, 8455.4mm^2 for 450kgf normal load.

By aid of the optical transducer the normal approach in the wheel-surface contact, was determined for the applied loading levels. Thus, for a normal load of 50kgf, a displacement of 2 mm was obtained. At 200kgf and 450kgf, normal approaches of 15mm and 24 mm were determined respectively.

A torque of 80Nm was also applied in the case of the 200kgf normal load. The effect of applying this torque on the shape, the contact area and its position can be visualized in Fig. 18.8. An increase in contact area from 4899.3mm^2 (only normal load applied) to 5205.8mm^2 (both normal load and torque applied) was observed. The applied torque induces a small shift in the position of the geometrical center of the contact patch. This leads to a small decrease in magnitude for the normal load of the contact. This is due to the occurrence of a small load component present in the

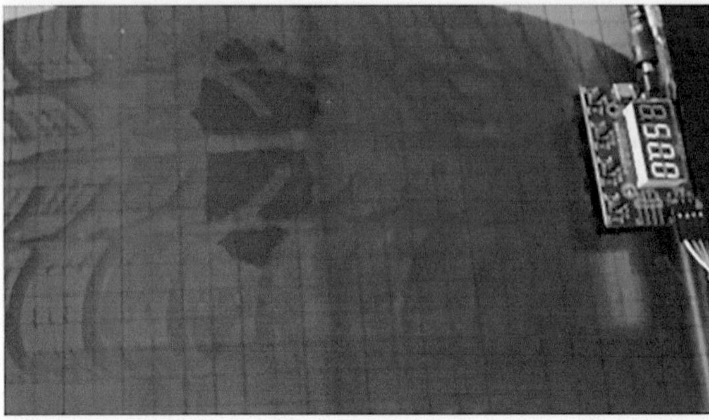

(a)

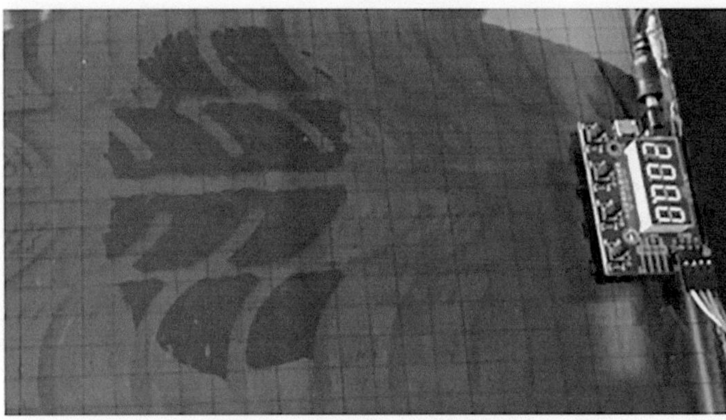

(b)

(c)

Fig. 18.6 Contact patch evolution for **a** 50kgf, **b** 200kgf, **c** 450kgf

Fig. 18.7 Partial detail of contact patch **a** and numerical results **b**, obtained with Image J software for a normal load of 50Kgf

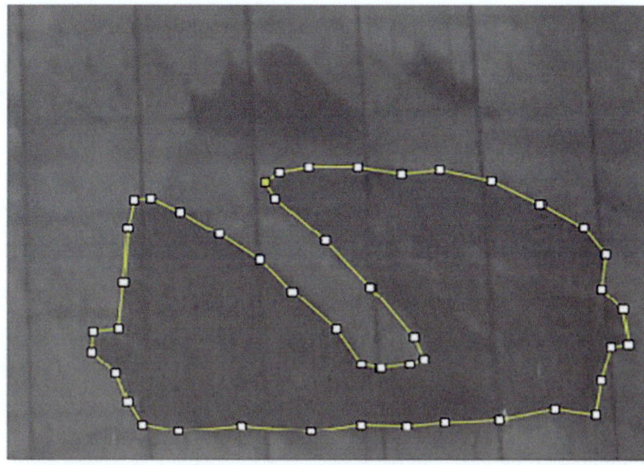

(a)

Results

File Edit Font Results

	Area	Mean	Min	Max
1	18.440	85.338	69	112
2	77.169	80.650	69.667	103.333
3	674.845	82.665	61.000	153.000
4	657.541	86.291	67.000	120.333
5	192.097	88.415	68.000	122.667

(b)

contact patch plane, tangent to the wheel surface. This component is responsible for the observed increase in contact patch area.

Fig. 18.8 Contact patch area for 200kgf normal load and 80Nm torque

18.6 Conclusions

An experimental set-up was designed and built to visualize the shape and size of the contact patch between a car tire and the road surface. The test rig uses a series of components taken from the real structure of an automobile in order to experimentally model real working conditions as rigorously as possible. The set-up is fitted with a series of transducers used to monitor and control the load level of the wheel-road contact, to measure the tire pressure, to determine the tire deformation and to control the torque transmitted by the planetary shaft. Thanks to these transducers, the experimental set-up allows determining the shape and size of the contact area in different situations obtained by varying the above-mentioned parameters.

In order to verify the proper functioning of the set-up, a FEM analysis was performed for the half-frame element of the structure. The analysis revealed that the maximum stresses in the half-frame are 256.7MPa, lower than the admissible ones for the material it is made of, and the maximum displacements on the loading direction are 5.99mm, manifesting on the end of the arm, which does not influence the good functioning of the test rig.

Preliminary experimental tests were carried out for a tire pressure of 2.35 bar, varying the wheel load level between 50 and 450Kgf. For each load level, the relative position of the planetary shaft with respect to the raceway was measured and recorded, as well as the shape and size of the contact patch area, respectively, following an image analysis with ImageJ software. The results showed that the contact patch area increased with increasing load level. The normal approach between the contact bodies also increased with the amount of normal loading.

Another test was carried out by adding an 80Nm torque transmitted to the wheel, while a normal load of 200kgf was applied. An increase in the contact patch and a displacement of the contact patch in the direction of wheel displacement imposed by the direction of applied moment was obtained.

References

1. Guo, M., Zhou, X.: Tire-pavement contact stress characteristics and critical slip ratio at multiple working conditions. Adv. Mater. Sci. Eng (2019)
2. Douglas, R., Woodward, W.D.H., Woodside, A.R.: Road contact stresses and forces under tires with low inflation pressure. Can. J. Civ. Eng. **27**(6), 1248–1258 (2000)
3. Guo, M., Liu, P., Feng, W., Li, X., Ge, H.: Analysis of tire contact stresses and asphalt pavement rutting under gradient temperature and typical driving conditions. J. Metrics. Sci. Rep **15** (2025)
4. Moisescu, R., Frățilă, G.: Finite element model of radial truck tyre for analysis of tyre—road contact stress. U.P.B. Sci. Bull **73**, 3 (2011)
5. Minca, C.: The determination and analysis of tire contact surface geometric parameters. Rev. Air Force Acad. **1**, 28 (2015)
6. Aguilar-Martínez, J., Alvarez-Icaza, L.: Analysis of tire-road contact area in a control oriented test bed for dynamic friction models. J. Appl. Res. Technol (13) (2015)
7. Saleh, M., Noordermeer, J.W.M., Reuvekamp, L.A.E.M., Blume, A.: Understanding test modalities of tire grip and laboratory road correlations with modeling. Tribology Letters (2021)

Part IV
Social Impact of the Technology and Industrial Management

Chapter 19
A Behavioral Finance Research Agenda in the Era of AI-Driven Commercial Banking

Nur Iylia Syafiqah Binti Abdul Malik and Anusuyah Subbarao

Abstract The economic sector and society are increasingly focusing on artificial intelligence (AI). AI was effectively implemented in the banking sector throughout its early stages. AI is predominantly employed in investment banking and backend operations that do not need direct customer touch. Thus far, the focus on client engagement has resulted in limited attention being given to the implementation of AI in the realm of commercial banking. The integration of AI in the commercial banking sector has the capacity to modify business protocols and client interactions, perhaps leading to the emergence of research prospects in the field of behavioral finance. To address this research gap, this study conducted a systematic review of existing literature to identify the various uses of AI in commercial banks, as well as the challenges involved with implementing AI in this context. Research findings suggest that commercial banks have the potential to enhance consumer targeting, minimize lending losses, enhance payment processing security, streamline compliance-related tasks, and expand their client base. Researcher has raised concerns about the implementation of technological advantages, the integration of AI into corporate operations, and ensuring user acceptability through transparency, privacy, and proper documentation.

Keywords Artificial intelligence (AI) · Commercial banking · Investment banking · Behavioral Finance

N. I. S. B. A. Malik (✉)
Faculty of Business, Multimedia University, Cyberjaya, Malaysia
e-mail: iylia.malik@mmu.edu.my

A. Subbarao
Faculty of Management, Multimedia University, Cyberjaya, Malaysia

R. Tamas and C. A. Heiman (eds.), *Selected Papers from AToMS 2024*,
Lecture Notes on Multidisciplinary Industrial Engineering,
https://doi.org/10.1007/978-3-032-02452-7_19

19.1 Introduction

Commercial banks serve increasingly complex needs, drastically in an ever-changing technological environment. The merging of digital transformation and AI can grasp the enormous potential and offers significant advantages in the commercial banking industry. AI-powered algorithms are beneficial for credit scoring, risk assessment, fraud detection, and customer service automation. This convergence fosters innovation, improves transaction processing efficiency, and ensures competitiveness in the evolving financial landscape. The study conducted by Tang & Tien [1] on the influence of AI on the Vietnam commercial bank system, based on a survey of 60 selected commercial banks, concluded that AI drastically accelerated bank's revenue. A recent study conducted by Alzyoud et. al [2] involving 198 personnel in the control and supervision departments of Jordanian commercial banks found that AI dimensions have a statistically significant impact on the efficiency of commercial banking services in Jordan.

Backend services are crucial for supporting banks' basic tasks. However, research indicates that incorporating AI in commercial banking can introduce security and privacy problems because of its heavy reliance on data, making it susceptible to breaches or misuse. The growing use of AI in commercial banking to automate operations may result in the replacement of human employees, thereby causing social and economic consequences [3].

As a result of a search conducted in the Emerald Web database as at January 2024, it has been discovered that a significant number of studies [4–6] continue to emphasize the benefits of AI in a few services offered by commercial banking such as investing area. Commercial banking is implementing AI-based smart solutions to assist customers in making investment decisions and to support their investment banking research. There has been a very limited amount of research conducted on the application of AI in backend services in commercial banking [7] for instance risk assessment, compliance monitoring, and fraud detection. These components are all necessary for the efficient operation of banking operations, but they may not directly entail interaction with customers.

The research objective that can be formulated for this study is to investigate the effects of integrating AI technologies on backend services in the commercial banking industry and to explore the consequences. Based on the gap, this research aims to focus on the lack of information in current literature about the use of AI in backend services in commercial banking, while also examining prospective outcomes. This study will accomplish the research objective by conducting a structured literature review through the utilization of the methodology described by Webster and Watson [8]. An examination of the benefits and challenges associated with applying AI in commercial banking is included in the literature review organized based on operations of commercial banks.

This paper will present the literature review, methodology, discussion and finally the conclusion.

19.2 Literature Review

19.2.1 Commercial Banks

Commercial bank referred to a financial institution that serves as the backbone of the financial system. These include supporting the functioning of businesses and fostering modern economic development as they enable business transactions, transfer funds from savers to borrowers, and stimulate economic development [2]. Among the everyday banking services provided by commercial banks are deposit taking, credit and loan provision, payment processing, investment product sales, and company financial management.

19.2.2 Evolution of AI in Commercial Banking

The roots of AI in banking can be traced back to the 1950s, when early computer scientists began exploring the concept of machine learning (ML) and its potential applications in the financial industry [9]. Over the decades, AI has evolved from basic pattern recognition and automation to more sophisticated applications. Indeed, AI is transforming the banking sector. The field of investment management and trading is one of the key early applications of AI in the banking industry. Banks leveraged AI algorithms to analyze market data and make investment decisions. Necessarily, their goal was to generate higher returns [4]. This allowed banks to process and act on vast amounts of financial data much faster than human analysts could [10] (Table 19.1).

Table 19.1 AI functions in commercial banking

Function	Description	Technologies/ Techniques	Source
Risk management	Enhancing banks' ability to detect financial crimes and suspicious transactions	Pattern Recognition, Machine Learning	[11, 12, 10]
	Robust and dynamic risk modeling to predict and mitigate market risks	Advanced Risk Modeling Algorithms	[14, 6, 13]
Back-office efficiency	Automating data extraction and analysis for processes like credit risk scoring and financial reporting	Computer Vision, Optical Character Recognition (OCR)	[12, 15, 3]
Customer service	Advanced AI applications, such as generative AI, to handle customer service tasks	Generative AI, AI-powered Assistants	[16]
	AI-powered assistants for natural, conversational interaction with customers	Natural Language Processing (NLP)	[1, 11, 7, 16, 17]

In a nutshell, the evolution of AI in banking has been a journey of continuous innovation. AI has evolved from the its first stage of automation and pattern recognition to the more recent advancements. The advancements includes areas such as risk management and customer experience. AI has become a critical component of the modern banking landscape.

19.2.3 Current State of AI in Commercial Banking

AI is being increasingly adopted in the Malaysian banking sector to enhance efficiency and customer service. According to a report by Accenture, 70% of Malaysian financial services executives strongly believe that AI and virtual assistants are the future of customer service [18]. Indirectly it shows that Malaysian banking consume are ready to embrace AI. AI adoption is now taking place in Malaysia according to a survey by Bank Negara Malaysia (BNM) on the use of AI/ML by financial service providers (FSPs) in the country. Their survey consist of 25 respondents including banks, insurers and payments operators [19].

Malaysian bank customers also have a strong appetite for tech innovations to enable digital banking services. A Unisys survey reveals that Malaysian bank customers are ready to embrace AI and automation in digital banking, depending on the type of service [20]. However, while traditional banking is making progress and incorporating novel technologies like AI, banks in Malaysia have not yet reached the point of an AI revolution, and the human touch is still crucial.

In summary, AI adoption in Malaysian commercial banking is in a preliminary stage, with banks exploring ways to integrate AI to enhance operational efficiency and customer service [20]. However, the full AI revolution is yet to come, and human involvement remains important in the banking sector.

19.2.4 Behavioral Finance Implications

The growing influence of artificial intelligence (AI) in the financial services industry has significant implications for the field of behavioral finance. This study highlights several key research opportunities at the intersection of AI and behavioral finance:

- Robo-advisors and behavioral financial decision-making: The increasing number of robo-advisors, automated investment platforms that use quantitative algorithms to manage investor portfolios. It presents an opportunity to study how AI-driven financial advice impacts the behavioral biases and decision-making processes of investors [21]. In the context of investment decisions, researchers can examine how robo-advisors influence factors like herding behavior, overconfidence, and loss aversion in investment [22, 23].

- AI personalization and framing effects: AI has the power and ability to customize financial recommendations and messaging. It is based on individual customer data opens the door to study the impact of different behavioral framing techniques [24]. This study found that simplifying messages and using negative framing can be more effective than positive framing in reducing overdraft fees and highlighting the importance of behavioral finance principles in AI-driven customer interactions.
- Ostrich effect: The research suggests that low-income individuals may be more prone to the "ostrich effect" [25]. They avoid checking their financial information. Instead of just reducing information costs but to effectively engage these customers and address their behavioral biases [26, 27].
- Integrating AI and behavioral finance in financial advisory: Since AI becomes common in financial advisory services, there is an opportunity to develop frameworks that optimally combine quantitative algorithms with an understanding of human behavioral tendencies [10]. Researchers can explore how to design AI-powered financial advice that accounts for and mitigates common behavioral biases.
- Ethical considerations: The increasing use of AI in financial decision-making raises important ethical questions. The questions are about transparency, accountability, and potential misuse of behavioral insights [20, 28]. Researchers should investigate the governance and oversight mechanisms needed to ensure AI applications in finance respect principles of fairness and consumer protection.

In future, the growing influence of AI in the financial industry presents a wide area for behavioral finance research. The possible areas include understanding how AI impacts individual decision-making to developing frameworks for responsible integration of AI and behavioral insights in financial services.

19.3 Methodology

This study was to summarize and combine the available literature. The methodology that was used is a narrative review. For the purpose of this research, articles and journals were chosen with a focus on AI-Driven Commercial Banking across diverse global settings. All the collected material is rigorously examined and synthesized. The data spans the most recent five years, beginning with publications from 2019 and ending with those from 2023, and is organized chronologically. The literature was gathered by authors to provide their individual evaluations of the findings, identifying trends, patterns, and gaps.

This methodology has several advantages. One of which is the engagement of seasoned professionals who have a thorough understanding of prior research. This research make a contribution by providing concise summaries, insightful interpretations and meaningful connections to the impactful studies. Examples of narrative reviews commonly used in financial research include those by Anshika & Singla (2022), Huston (2010), and Lusardi (2019).

19.4 Discussion

19.4.1 Applications of AI in Commercial Banking

See the Table 19.2

19.4.2 AI's Impact on Customer Behavior

AI is transforming how businesses interact with customers and influence purchasing decisions [17]. Personalized experiences by AI's predictive power can boost customer satisfaction and engagement [33]. However, as these AI-powered services become the norm, businesses must continually assess and address evolving customer needs to remain competitive.

The benefits of AI personalization must be balanced with consumer concerns over data privacy [28]. Businesses that prioritize data privacy and transparency tend to earn greater customer loyalty. Ethical considerations around AI's societal impact are also becoming a key factor in purchasing decisions, as ethical consumers favor companies committed to responsible AI practices.

19.4.3 Behavioral Insights from AI Data

AI provides new opportunities into financial behaviors, risk preferences, and consumer psychology. By analyzing vast datasets, AI can uncover previously hidden patterns and trends in consumer decision-making [17]. This could lead to more accurate models of financial behaviors and better-informed strategies for nudging consumers towards beneficial financial decisions.

19.4.4 Emerging Research Areas

Potential research directions at the intersection of AI and behavioral finance include:

- Investigating how AI-powered personalization and recommendation systems impact consumer financial decision-making and risk-taking [15]

Table 19.2 Applications of Ai in commercial banking

Application Area	Description	Technologies/ Techniques	Example	Impact/ Benefits
Consumer targeting	AI analyzes personal financial data, purchasing behavior, and social media activity for real-time customer segmentation and personalized marketing	Data Analytics, Machine Learning	Maybank's Maya: Analyzes customer data to recommend tailored investment products and credit card offers	Increased engagement, enhanced conversion rates, improved satisfaction
Lending processes	AI improves credit scoring, risk assessment, and automated underwriting by analyzing vast customer data, enhancing risk predictions	Predictive Modeling, Automated Underwriting	CIMB e-Zi Lending Platform: Processes personal loan and SME financing applications using AI for credit scoring and risk assessment	Faster approvals, accurate risk assessment, increased efficiency
Payment processing security	AI monitors transactions in real-time, detecting anomalies and potential fraud, allowing for immediate intervention to prevent losses	Real-Time Monitoring, Fraud Detection	RHB's Fraud Detection System: Monitors credit card and online banking transactions for suspicious activity	Enhanced fraud detection, rapid response to threats, increased customer trust
Compliance and regulatory tasks	AI automates Know Your Customer (KYC) and Anti-Money Laundering (AML) processes, streamlining compliance functions, reducing errors, and improving adherence to regulations	Automation, AI Algorithms	Public Bank's Compliance Automation System: Automates KYC and AML processes, generates compliance reports, and screens for sanctions	Increased efficiency, enhanced accuracy, regulatory adherence
Client engagement and expansion	AI tools like virtual chatbots provide 24/7 support and personalized interactions, enhancing customer experience and retention	AI Chatbots, Predictive Analytics, Recommendation Engines, Customer Segmentation	CIMB's AI-Driven Customer Engagement Platform: "Eva" virtual assistant, personalized product recommendations, targeted marketing	Improved customer support, proactive services, reduced churn

- Exploring the role of AI-generated content and virtual assistants in shaping consumer financial behaviors and perceptions [9]
- Analyzing the ethical implications of using AI for financial decision-making, including issues of bias, transparency, and consumer trust
- Developing AI-driven models to predict and influence consumer financial behaviors, such as savings, investments, and debt management.

By leveraging the power of AI to gain novel behavioral insights, researchers can uncover new opportunities to enhance the customer experience in the banking and finance sectors.

19.5 Conclusion

AI applications in commercial banking are transforming operational efficiency and customer interactions. By integrating advanced technologies, banks can enhance consumer targeting, streamline lending processes, bolster payment security, and improve compliance and regulatory tasks.

In future research, examining AI's impact on customer behavior and decision-making is crucial. This includes how AI personalization influences financial choices and addressing ethical concerns. Exploring AI's role in shaping consumer behavior and developing predictive models for financial decisions can provide deeper insights into enhancing customer experiences and fostering responsible AI use in banking.

Acknowledgements The authors are also grateful to Multimedia University (MMU) Malaysia for the financial support.

References

1. Tang, S.M., Tien, H.N.: Impact of artificial intelligence on Vietnam commercial bank operations. Int. J. Soc. Sci. Econ. Invent. **6**(07), 296–303 (2020). https://doi.org/10.23958/ijssei/vol06-i07/216
2. Alzyoud, A.S., Al-Naimi, S., Al-Gasaymeh, A.A.: The impact of artifial intelligence on the efficiency of banking service: evidence from Jordanian commercial banks. J. Southwest Jiaotong Univ. **57**(6), 73–90 (2022)
3. Mehta, V.: The Impact of AI on the banking industry. Mutualmobile.Com, vol. 1, no. 1 [Online] (2023). https://mutualmobile.com/blog/the-golden-ticket-ai-in-the-chocolate-industry
4. Ashta, A., Herrmann, H.: Artificial intelligence and fintech: an overview of opportunities and risks for banking, investments, and microfinance. Strateg. Chang. **30**(3), 211–222 (2021). https://doi.org/10.1002/jsc.2404
5. Saba, C.S., Pretorius, M.: The impact of artificial intelligence (AI) investment on human well-being in G-7 countries: Does the moderating role of governance matter? Sustain. Futur. **7**, 100156 (2024). https://doi.org/10.1016/j.sftr.2024.100156

6. Toorajipour, R., Sohrabpour, V., Nazarpour, A., Oghazi, P., Fischl, M.: Artificial intelligence in supply chain management: a systematic literature review. J. Bus. Res. **122**, 502–517 (2021). https://doi.org/10.1016/j.jbusres.2020.09.009
7. McKinsey & Company.: Building TheBank of The Future. Glob. Bank. Pract. [Online] (2021). https://www.home.barclays/content/dam/barclayspublic/docs/InvestorRelations/AnnualReports/AR2016/BarclaysPLCAnnualReport2016.pdf
8. Webster, J., Watson, R.T.: Analyzing the past to prepare for the future: writing a literature review. MIS Q. 26(2), xiii–xxiii (2002). 10.1.1.104.6570
9. Gupta, P.: Leveraging machine learning and artificial intelligence for fraud prevention. Int. J. Comput. Sci. Eng. **10**(5), 47–52 (2023). https://doi.org/10.14445/23488387/ijcse-v10i5p107
10. Atadoga, A., Obi, OC., Onwusinkwue, S., Dawodu, S.O., Osasona, F., Daraojimba, A.I.: AI's evolving impact in US banking: An insightful review. Int. J. Sci. Res. Arch. **11**(1), 904–922 (2024). https://doi.org/10.30574/ijsra.2024.11.1.0157
11. Mhlanga, D.: Industry 4.0 in finance: the impact of artificial intelligence (AI) on digital financial inclusion. Int. J. Financ. Stud. **8**(3), 1–14 (2020). https://doi.org/10.3390/ijfs8030045
12. Baviskar, D., Ahirrao, S., Potdar, V., Kotecha, K.: Efficient automated processing of the unstructured documents using artificial intelligence: a systematic literature review and future directions. IEEE Access **9**, 72894–72936 (2021). https://doi.org/10.1109/ACCESS.2021.3072900
13. Qin, J., Qin, Q.: Cloud platform for enterprise financial budget management based on artificial intelligence. Wirel. Commun. Mob. Comput. **2021** (2021) https://doi.org/10.1155/2021/8038433
14. Cao, L., Yuan, G., Leung, T., Zhang, W.: Special issue on AI and FinTech: The Challenge Ahead. IEEE Intell. Syst **35**(3), 3–6 (2020). https://doi.org/10.1109/MIS.2020.2983636
15. Pandey, M.K., Sergeeva, I.: Artificial intelligence impact evaluation: transforming paradigms in financial institutions. World Econ. Manag. **22**(1), 147–164 (2022). https://doi.org/10.25205/2542-0429-2022-22-1-147-164
16. Talib, A., Housni, M., Radid, M.: Utilizing M-technologies for AI-driven career guidance in morocco: an innovative mobile approach. Int. J. Interact. Mob. Technol. **17**(24), 173–188 (2023). https://doi.org/10.3991/IJIM.V17I24.44263
17. Biswas, K., Patra, D.G.: Role of artificial intelligence (AI) in changing consumer buying behavior. Int. J. Res. Publ. Rev. **04**(02), 943–951. https://doi.org/10.55248/gengpi.2023.4227
18. Accenture.: Driving reinvention, delivering 360° value Annual Report (2023)
19. Bank Negara Malaysia.: Artificial intelligence in the Malaysian Financial System: Opportunities, Risks, and the Way Forward, pp. 27–31 (2023)
20. Husain, A.R.A.M., Hamdan, A., Fadhul, S.M.: The impact of artificial intelligence on the banking industry performance. Stud. Comput. Intell. **1037**(1), 145–156 (2022). https://doi.org/10.1007/978-3-030-99000-8_8
21. Dietzmann, C., Jaeggi, T., Alt, R.: Implications of AI-based robo-advisory for private banking investment advisory. J. Electron. Bus. Digit. Econ. **2**(1), 3–23 (2023). https://doi.org/10.1108/jebde-09-2022-0037
22. Altrock, S., Mention, A.L., Aas, T.H.: Being human in the digitally enabled workplace: insights from the robo-advice literature. IEEE Trans. Eng. Manag 1–16 (2023). https://doi.org/10.1109/TEM.2023.3291820
23. Manolova, T.S., Brush, C.G., Edelman, L.F., Elam, A.: Pivoting to stay the course: How women entrepreneurs take advantage of opportunities created by the COVID-19 pandemic. Int. Small Bus. J. Res. Entrep. **38**(6), 481–491 (2020). https://doi.org/10.1177/0266242620949136
24. Weixiang, S., Qamruzzaman, M., Rui, W., Kler, R.: An empirical assessment of financial literacy and behavioral biases on investment decision: fresh evidence from small investor perception. Front. Psychol. **13**(September), 1–18 (2022). https://doi.org/10.3389/fpsyg.2022.977444
25. Prosad, J.M., Kapoor, S., Sengupta, J.: Theory of behavioral finance. Handb. Res. Behav. Financ. Invest. Strateg. Decis. Mak. Financ. 1–24 (2015). https://doi.org/10.4018/978-1-4666-7484-4.ch001

26. Sharma, A., Kumar, A.: A review paper on behavioral finance: study of emerging trends. Qual. Res. Financ. Mark. **12**(2), 137–157 (2020). https://doi.org/10.1108/QRFM-06-2017-0050
27. Kute, D.V., Pradhan, B., Shukla, N., Alamri, A.: Deep learning and explainable artificial intelligence techniques applied for detecting money laundering-A critical review. IEEE Access **9**, 82300–82317 (2021). https://doi.org/10.1109/ACCESS.2021.3086230
28. Berendt, B., Littlejohn, A., Blakemore, M.: AI in education: learner choice and fundamental rights. Learn. Media Technol 312–324 (2020). https://doi.org/10.1080/17439884.2020.1786399

Chapter 20
Navigating AI Acceptance in the Restaurant Realm: Assessing the Potential Replacement of Humans by AI in the F&B Industry

Anusuyah Subbarao, Dragos Taralunga, and Nur Iylia Syafiqah Abdul Malik

Abstract In recent years, the Food and Beverage (F&B) industry has undergone a transformative integration of Artificial Intelligence (AI) devices, marking a paradigm shift in its operational landscape. The incentive behind this widespread adoption of AI devices stems from a confluence of factors, with the global pandemic catalyzing change. Faced with unprecedented challenges, businesses in the F&B sector have turned to AI solutions to navigate disruptions and to ensure the continuity of operations while also maintaining proper social distancing protocols. This surge in AI utilization in the F&B industry has prompted the present comprehensive study that examines the intricate dynamics of AI acceptance. The research focuses on key determinants such as perceived performance expectancy, perceived effort expectancy, and perceived anthropomorphic characteristics. While previous investigations have extensively explored AI adoption in various sectors, the F&B industry presents unique challenges and opportunities. The study not only seeks to explain the factors influencing AI acceptance but also explores consumer perceptions of these technological implementations within the F&B realm. This study has been conducted using a quantitative method using purposive sampling. The target population of this study was in the Klang Valley area, Malaysia with a sample size of 107. Data was analyzed using the IBM Statistical Package for the Social Sciences (SPSS) v29. Findings

A. Subbarao (✉)
Department of Information Technology Faculty of Management, Multimedia University, Cyberjaya, Malaysia
e-mail: anusuyah.subbarao@mmu.edu.my

D. Taralunga
Faculty of Electronics, Telecommunications and Information Technology, National University of Science and Technology Politehnica Bucharest, Bucharest, Romania
e-mail: dragos.taralunga@upb.ro

N. I. S. A. Malik
Faculty of Business, Multimedia University, Malacca, Malaysia
e-mail: iylia.malik@mmu.edu.my

R. Tamas and C. A. Heiman (eds.), *Selected Papers from AToMS 2024*,
Lecture Notes on Multidisciplinary Industrial Engineering,
https://doi.org/10.1007/978-3-032-02452-7_20

show that both positive and negative attitudes towards the acceptance of AI do occur depending on certain factors. When focusing on whether AI can improve the quality of service or possess more human-like features, people's opinions towards accepting AI devices sway in a more positive direction.

Keywords Acceptance of AI · F&B industry · Performance expectancy · Effort expectancy · Anthropomorphic characteristics

20.1 Introduction

During the pandemic, many people have lost their jobs, especially those who work in the service industry. Even with the pandemic slowing down, many industries that need close contact between people have begun to invest in AI approaches to replace their human employees, either to limit the spread of the virus or as an investment for the future. According to a study published by St John's University, the rise in digital transformation seen in companies is due to the coronavirus. Many traditional businesses will soon favor companies that adopt a more digital model such as using AI or switching to selling online [1].

It should be noted that the usage of AI in the restaurant industry has only emerged recently. AI gained fame when they were introduced as a way to replace human workers in manufacturing industries to boost productivity and limit human error [2]. AI allowed factories to operate on a larger scale with more precision than compared to traditional methods of manufacturing. During the beginning of the pandemic, many restaurants suffered from the social distancing guidelines that were imposed to limit the spread of the virus. Based on data analysis from OpenTable and Allen Aassociates [3], it was shown that during the height of the pandemic, 89% of full-contact restaurants saw a 100% decrease in their traffic. As the restaurant industry is a high-contact service industry, many were forced to shut down to limit the spread of the virus, causing many restaurant owners and workers to lose their jobs. This led to restaurants having to find ways to operate their businesses as usual all while keeping up with social distancing guidelines. Thus, the usage of AI in restaurants saw a big boom during this time.

However, AI is still in its infancy and its future within the restaurant industry is still uncertain. Some experts have argued that no matter how advanced AI can be, it would still be unable to replace human workers. This is due to the restaurant industry being under the service industry. In many cases, a high-contact industry performs better when its staff exhibits high emotional and social skills, something that AI has failed to replicate. In a recent study on building robots for quick-service restaurants, many of its participants gave feedback on how the robots would be more well-received if they had more human-like characteristics in terms of appearance, voice, and ways of interacting with customers [4]. This proves that while AI does provide many benefits, humans prefer more organic interactions when receiving services. On the other hand, the usage of AI staff has also increased due to the pandemic. The coronavirus has caused many customers now to have a slight obsession with a restaurant's overall cleanliness. This led many to prefer robot staff over humans, as robots have a lower risk of spreading germs to food and customers. [5]. With two angles in this story, this poses a question: will AI eventually replace human workers in the restaurant industry? Or will humans prevail?

With technology evolving at a rapid rate, implementing AI into the restaurant industry is a good investment for the future. The restaurant industry is also tied to the tourism industry; and with tourism booming after the pandemic, having something special to lure customers in can boost sales. For example, self-service technologies, mobile apps, and other forms of AI can help increase the overall number of customers and their customer satisfaction [6]. Some experts also mentioned that there would be a rise in AI usage within the service industry as opposed to the traditional style of service [7]. Compared to other industries, the research on whether the rise of AI acceptance will cause humans to be replaced by AI devices in the restaurant industry is lesser. As such, this research will investigate whether AI will replace human workers in the restaurant industry in Malaysia.

Moving forward, the goal of this research is to find the relationship between perceived performance expectancy, perceived effort expectancy and perceived anthropomorphic characteristics with the acceptance of the use of AI devices in the restaurant industry through the mediating variable of emotion.

This paper will present the related work, methodology, results, findings, and finally the conclusion.

20.2 Related Work

20.2.1 Dependent Variable- the Acceptance of AI Devices in the Restaurant Industry

The usage of AI in industries has increased in the past few years. These numbers are expected to rise due to the impact of Covid-19 [8]. The Coronavirus has made many people prioritize health and cleanliness, especially in the restaurant industry. Cross-contamination and maintaining general cleanliness in restaurants are some things that most consumers look out for these days, which prompted many restaurants to implement AI devices to limit human interaction.

Although it has been quite some time since QR menus have been implemented in restaurants, many continue to ask for the abolishment of the system. The Washington Post even made an article criticizing the QR menus, going so far as to say that "QR menus are the death of civilization" [9]. For instance, in sunny weather, a human server could suggest a cold beverage to a customer rather than a hot one. An AI would not be able to do so as most systems are not programmed to give suggestions to customers based on the daily weather report. This lack of emotion and spontaneity in AI can be seen as a big factor in why AI may not be able to replace human workers in the restaurant industry.

20.2.2 Independent Variable

20.2.2.1 Performance Expectancy

Performance expectancy can be defined as the extent to which one believes that using a particular system can help them improve their performance while doing their tasks. This construct is included within a theory called the Unified Theory of Acceptance and Use of Technology (UTAUT) which was developed to measure user acceptance and continued usage of a system [10]. In this study, performance expectancy will be used to measure whether individuals have a positive emotion toward using AI in the restaurant industry.

Performance expectancy is key to understanding whether AI will replace human workers, especially with its increased use in restaurants after COVID-19. AI robots are generally seen as being cleaner than humans as robots can be programmed to always self-disinfect, avoid cross-contamination, and follow food safety protocols [11]. On the other hand, humans can sometimes neglect to follow safety protocols which can increase the spread of viruses. This may cause people to develop a fondness for having AI robots in restaurants more so than human workers. If individuals see AI as improving restaurant operations, they will be more accepting of it, increasing the likelihood that AI will replace human workers in the future.

20.2.2.2 Effort Expectancy

Effort expectancy is defined as the extent to which an individual thinks that a technology will be easy to use. This construct is also part of the Unified Theory of Acceptance and Use of Technology (UTAUT) [10]. Effort expectancy is an important variable in this research as it has to be used to measure whether people hold a positive or negative image of using AI in restaurants. Although AI does have its benefits, people are usually daunted by having to use certain technologies due to the complexity of the systems. Most restaurants do not include instructions on how to use the AI devices such as self-ordering kiosks and tablets which can send people into confusion when trying to order simple things. The lack of clarity and help when operating these AI devices will create a negative image for the usage of AI devices. Therefore, this can increase the objection to having AI in restaurants. This rings especially true to those of the older generations.

Studies have shown that people of the older generation are hesitant to use most forms of technology as they negatively perceive it as too complicated to use and due to the lack of instructions on its use [12]. The younger generation has also expressed dissatisfaction with using AI in restaurants for similar reasons, which shows that there is a large group of individuals out there that does not approve of AI usage in the restaurant industry. Therefore, if people find AI too complex to use, they will be less likely to accept it in restaurants, reducing the possibility of AI replacing human workers in the industry.

20.2.2.3 Anthropomorphic Characteristics

According to the Cambridge dictionary, anthropomorphism is defined as the phenomena of treating non-human things, such as animals and objects, like humans by associating them with human-like characteristics. These characteristics can range anywhere from physical characteristics (e.g., human-like eyes, nose, lips) to mental characteristics such as emotional capacity (the ability to feel and interpret human emotions) [13]. There have been various studies on the importance of anthropomorphism in AI and it has been confirmed that most people are willing to interact with AI but under certain conditions. The biggest criterion is how human the AIs can be. For example, a study made on which AI chatbot is more favorable to customers found that most customers are more responsive and willing to continue engaging in conversation with the AI that seemed to have a more human-like way of communicating [14].

Another study that also used AI chatbots in their research found that most respondents were disappointed with the AI chatbots as they lacked human emotions during communications [15]. This lack of perceiving human emotions also resulted in the AI's inability to solve more complex or specific problems which caused further dissatisfaction with the respondents. This means that if individuals perceive AI's lack of human empathy and emotion to be a barrier, they would be less accepting of having AI in restaurants. This would reduce the chance of AI replacing human workers in restaurants.

20.2.3 Mediating Variable- Emotion

Emotion in the context of this study pertains to how people feel about the usage of AI in the restaurant industry. The respondents will assess their willingness or objection to using AI devices in the restaurant industry based on the advantages and disadvantages of using said AI devices. Respondents will decide how they feel about the usage of AI devices in restaurants based on the previous three variables: perceived performance expectancy, perceived effort expectancy, and perceived anthropomorphic characteristics [10]. This study suggests that these three variables have a higher probability of being used to assess the advantages and disadvantages of utilizing AI devices in restaurants.

As such, these three variables might have a significant influence on the emotions of people towards AI devices which can ultimately determine whether or not this technology will be able to fully replace human workers in restaurants in the foreseeable future. It is hypothesized that stronger negative emotions towards using AI devices can be influenced by higher perceived effort, lower perceived performance, and less perceived anthropomorphic characteristics.

This hypothesis is further supported by the Cognitive Dissonance Theory wherein it proposes that humans will rarely go against their preconceived views and opinions to avoid feeling uncomfortable [16]. This means that when people have decided how they feel about a certain situation, object, or person, they will not change their minds easily. This makes it easier to assess the respondents' opinions on AI usage in restaurants produced.

20.3 Methodology

The theoretical framework shown in Fig. 20.1 is proposed based on the AI Device Use Acceptance (AIDUA) model by Gursoy et al. [17] and the Consumer-AI device relation adapted from Pelau et al. [13].

Based on this, the hypothesis is shown in Table 20.1.

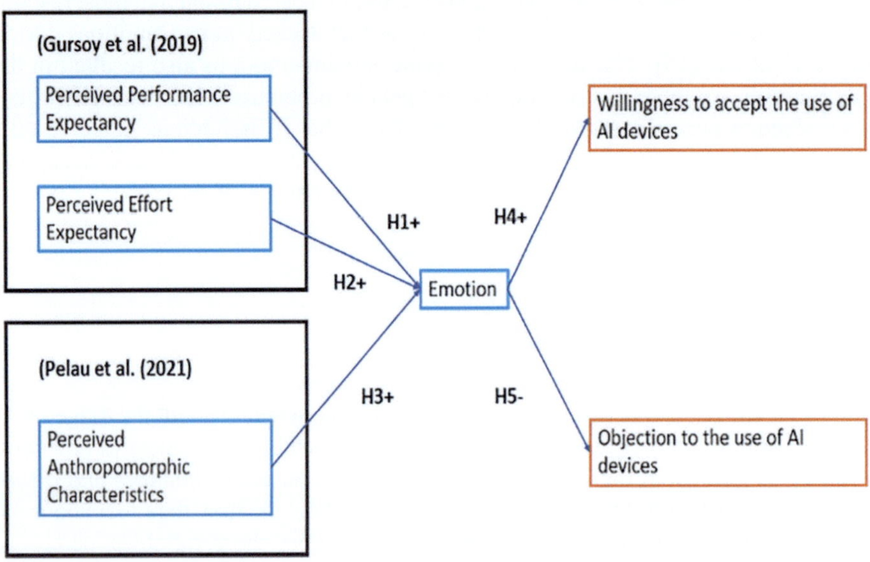

Fig. 20.1 Research theoretical framework

Table 20.1 List of hypothesis

	Hypothesis
H1	Perceived performance expectancy will generate a positive relationship with positive emotions toward the use of AI devices
H2	Perceived effort expectancy will generate a negative relationship with positive emotions toward the use of AI devices
H3	Perceived anthropomorphism will generate a positive relationship with positive emotions toward the use of AI devices
H4	Emotion has a positive relationship with willingness to accept the use of AI devices
H5	Emotion has a negative relationship with the objection to the use of AI devices

20.4 Results and Findings

There is a positive relationship between performance expectancy and the acceptance of AI in the restaurant industry. Many respondents have a positive attitude toward accepting AI in the restaurant as they believe that machines are more accurate than human beings. As long as AI devices can improve the overall experience in the restaurant, humans will be more accepting and welcoming towards AI devices. They are willing to learn to use AI devices if that means human error can be reduced [18]. This can result in AI devices dominating the restaurant industry and replacing human workers.

There is a positive relationship between effort expectancy and the acceptance of AI in the restaurant industry. Although people are willing to learn to adapt to using AI devices in restaurants, doubts about the system's simplicity still cause respondents to have negative emotions towards accepting AI in the restaurant industry. If the AI devices prove to be too complicated to use, many people will begin to oppose using the devices in the restaurant. If the devices used are too high-tech, this might affect the ease of use and cause many to be unable to operate the devices, especially those of the older generation [19]. That may cause those of the older generation to not frequent restaurants that have AI devices and cause the restaurants to lose out on potential customers.

There is a slight positive relationship between anthropomorphic characteristics and the acceptance of AI in the restaurant industry. Although human empathy and characteristics are desired in an AI device, most respondents tend to care more about the performance and effort expectancy of AI devices and their emotions towards using AI devices in restaurants. Although anthropomorphic characteristics can influence people to more easily accept the usage of AI devices in the restaurant industry, it is not the most significant factor [13]. Many are willing to overlook the lack of humanoid characteristics in favor of improving the quality of service. Therefore, anthropomorphic characteristics can increase the acceptance of AI usage in the restaurant industry.

Emotion does mediate people's perceptions and the acceptance of AI in restaurants. In terms of willingness to accept and the objection to the use of AI devices in the restaurant industry, it is proven from the analysis to have a positive relationship towards both outcomes. From the analysis, we can conclude that people's perceptions can foster both positive and negative emotions toward AI devices in the restaurant industry. This is because, no matter how long humans have adapted to using modern technology and no matter how tech-savvy an individual is, there still exists the fear of not being able to understand and use a new technology [20]. At the same time, people realize the benefits of having and using AI in the restaurant industry such as an increase in service quality and experience. Thus, this has led there to both willingness and objection to accepting AI devices in the restaurant industry.

20.5 Conclusion

This study identified the relationship between perceived performance expectancy, perceived effort expectancy, and perceived anthropomorphic characteristics with the acceptance of the use of AI devices in the restaurant industry. For emotions, it is shown that both positive and negative attitudes towards the acceptance of AI do occur depending on the factors that are focused on. When focusing on whether or not AI can improve the quality of service or possess more human-like features, the opinions of people towards accepting AI devices sway in a more positive direction. Conversely, when people focus on their doubts and worries about the simplicity of the devices and the ease of use, the acceptability of AI device usage sways in a more negative direction.

Future studies should assess more and different variables that may determine the acceptability of AI usage in the restaurant industry. This study should also not only be limited to restaurants and restaurant-goers in Malaysia but also to those outside of the country so light can be shed on people's opinions on the acceptance of AI usage in restaurants.

Acknowledgements The authors are also grateful to Multimedia University (MMU) Malaysia for the financial support. This work was also supported by a grant from the National Program for Research of the National Association of Technical Universities - GNAC ARUT 2023.

References

1. Li, S.: St John's University—Review of Business Volume 41, Number 1. Review of Business, Volume 41, Number 1 (2021). https://usiena-air.unisi.it/retrieve/handle/11365/1180047/445029/Review-of-Business-41%281%29-Jan-2021.pdf
2. Yang, L., Henthorne, T.L., George, B.: Artificial intelligence and robotics technology in the hospitality industry: current applications and future trends. In: George, B., Paul, J. (eds.) Digital Transformation in Business and Society. Palgrave Macmillan, Cham (2020). https://doi.org/10.1007/978-3-030-08277-2_13

3. Allen, A.: How Coronavirus is Impacting the Restaurant Industry. Aaron Allen & Associates, Global Restaurant Consultants (2020). https://aaronallen.com/blog/restaurant-industry-coronavirus
4. Zemke, D.M.T., Jason & Raab, Carola & Kim, Jungsun.: How To Build a Better Robot For Quick-Service Restaurants. J Hospitality Tourism Research 44. 109634802094638 (2020). https://doi.org/10.1177/1096348020946383. https://www.researchgate.net/publication/343469607_How_To_Build_a_Better_Robot_For_Quick-Service_Restaurants
5. Choi, S.: The rise of Service Robots in the hospitality industry: Some actionable insights. Boston Hospitality Review The Rise of Service Robots in the Hospitality Industry Some Actionable Insights Comments (2021). https://www.bu.edu/bhr/2021/10/04/the-rise-of-service-robots-in-the-hospitality-industry-some-actionable-insights/
6. Giroux, M., Kim, J., Lee, J.C., et al.: Artificial intelligence and declined guilt: retailing morality comparison between human and AI. J. Bus. Ethics **178**, 1027–1041 (2022). https://doi.org/10.1007/s10551-022-05056-7
7. Blöcher, K., Alt, R.: AI and robotics in the European restaurant sector: assessing potentials for process innovation in a high-contact service industry. Electron Markets 31, 529–551 (2021). https://link.springer.com/article/https://doi.org/10.1007/s12525-020-00443-2#Tab1
8. Reynolds, K.: There was an increase in AI during the pandemic. soon we won't be able to imagine life without it. World Economic Forum (2021). https://www.weforum.org/agenda/2021/02/covid-19-increased-use-of-ai-here-s-why-its-here-to-stay/
9. Olen, H.: Opinion | QR code menus are the death of Civilization. The Washington Post (2022). https://www.washingtonpost.com/opinions/2022/06/15/against-qr-code-restaurant-menus/
10. Venkatesh, V., Morris, M.G., Davis, G.B., Davis, F.D.: User acceptance of information technology: toward a unified view. MIS Quarterly, **27**(3), 425–478. https://www.jstor.org/stable/30036540?read-now=1&oauth_data=eyJlbWFpbCI6IjExOTExMDA2NTdAc3R1ZGVudC5tbXUuZWR1Lm15IiwiaW5zdGl0dXRpb25JZHMiOltdfQ&seq=1#page_scan_tab_contents
11. Qian, C., Murphy, S.I., Orsi, R.H., Wiedmann, M.: How can AI help improve food safety? Annu. Rev. Food Sci. Technol. **14**, 517–538 (2023)
12. Heaggans, R. C.: The 60's are the new 20's: Teaching older adults technology – ed (2012). https://files.eric.ed.gov/fulltext/EJ990630.pdf
13. Pelau, C., Dabija, D.C., Ene, I.: What makes an AI device human-like? The role of interaction quality, empathy and perceived psychological anthropomorphic characteristics in the acceptance of artificial intelligence in the service industry. Comput. Hum. Behav. **122**, 106855 (2021)
14. Hill, J., Ford, W.R., Farreras, I.G.: Real conversations with artificial intelligence: a comparison between human-human online conversations and human–chatbot conversations. Comput. Hum. Behav. **49**, 245–250 (2015)
15. Nguyen, T.: Potential effects of chatbot technology on customer support: a case study (2019)
16. Festinger, L.: Cognitive dissonance. Sci. Am. **207**(4), 93–106 (1962)
17. Adeniran, A.O.: Application of Likert scale's type and Cronbach's alpha analysis in an airport perception study. Scholar J Applied Sci Res **2**(4), 1–5 (2019)
18. Obiorah, M.G.S., Piper, A.M.M., Horn, M.: Designing AACs for people with aphasia dining in restaurants. In: Proceedings of the 2021 CHI Conference on Human Factors in Computing Systems, pp. 1–14 (2021)
19. Kwak, M.K., Lee, J., Cha, S.S.: Senior consumer motivations and perceived value of robot service restaurants in Korea. Sustainability **13**(5), 2755 (2021)
20. Manh-Tung Ho, Unar, J.A., Adikari, A., Okumus, B., Taherdoost, H., Muk, A., Dutot, V., Hope, A. L.B., Aydin, B., Kasilingam, D.L., Wang, Z., Chicca, J., Vuong, Q.-H., Kim, T.-J., Tømte, C., Oracle.: Rethinking technological acceptance in the age of emotional AI: Surveying gen Z (Zoomer) attitudes toward non-conscious data collection. Technology in Society (2022). https://www.sciencedirect.com/science/article/pii/S0160791X2200152X#sec6

Chapter 21
Unlocking Frontiers: Innovations and Advancements in Measurement and Simulation Techniques Shaping Today's Global Research Landscape

Angelica Morales Baylon

Abstract The keynote address introduces the topic and its significance in the context of modern research and innovation. An overview of global research across various disciplines is provided, highlighting the challenges and opportunities researchers face in pushing the boundaries of knowledge. The importance of cutting-edge measurement techniques in capturing data accurately and precisely is discussed, showcasing examples of how advancements in measurement technology have revolutionized research in different fields. The significance of simulation techniques in modeling complex systems and scenarios is explained, illustrating how simulations enable researchers to explore new frontiers and test hypotheses in a virtual environment. How innovation in measurement and simulation techniques are shaping the global research landscape is explored, highlighting examples of breakthroughs and advancements made possible through these technologies. The synergies between measurement and simulation techniques in driving innovation and discovery are emphasized, and the discussion of how combining these approaches can lead to a deeper understanding of phenomena and accelerate scientific progress is discussed. The challenges researchers may face in adopting and implementing new measurement and simulation techniques are addressed by discussing the potential opportunities for collaboration and interdisciplinary research in this evolving landscape. The keynote address ends by encouraging researchers to embrace innovation, collaboration, and creativity in their work and inspire everyone to continue pushing the boundaries of knowledge and exploring new frontiers in research.

Keywords Maritime research · Simulation techniques

A. M. Baylon (✉)
External Relations Office/PAEPI–Global Maritime, Academy of Asia and the Pacific, Mariveles, Bataan, Philippines
e-mail: ambaylon@maap.edu.ph

R. Tamas and C. A. Heiman (eds.), *Selected Papers from AToMS 2024*,
Lecture Notes on Multidisciplinary Industrial Engineering,
https://doi.org/10.1007/978-3-032-02452-7_21

21.1 Introduction

In today's dynamic research landscape, advancements in measurement and simulation techniques have revolutionized the way we understand and optimize complex systems, leading to ground-breaking developments in various fields. In the maritime industry, integrating cutting-edge measurement and simulation technologies has significantly enhanced our ability to model, analyze, and improve marine systems, thereby driving innovation and efficiency in vessel design, operation, and safety. Let us delve deeper into how these advancements are shaping the future of maritime research and innovation.

21.2 Challenges and Opportunities in Maritime Research

A general overview of the state of global research across various disciplines and the maritime sector, along with some common challenges and opportunities that researchers typically face in pushing the boundaries of knowledge, are provided.

21.2.1 Interdisciplinary Collaboration

There is a growing trend towards interdisciplinary research, where researchers from different disciplines come together to address complex challenges that require a multifaceted approach.

21.2.2 Open Access and Data Sharing

The movement towards open-access publishing and data sharing is gaining momentum. These practices make research findings more accessible to a broader audience and promote transparency in the research process.

21.2.3 Technological Advancements

Technological advances, such as artificial intelligence, machine learning, and big data analytics (Sim et.al, 2023), are transforming how research is conducted in various fields. They enable researchers to analyze vast amounts of data and make new discoveries.

21.2.4 Global Challenges

Researchers are increasingly focusing on pressing global challenges, such as climate change, public health crises, sustainable development, and social inequality, seeking innovative solutions to address these complex issues.

21.2.5 Funding and Resources

Securing funding for research projects remains a significant challenge for many researchers, particularly in emerging fields or areas with limited resources. Competition for grants and funding opportunities is often fierce.

21.3 Challenges and Opportunities in Maritime Research

21.3.1 Environmental Sustainability

Researchers in the maritime sector are working to develop sustainable practices to reduce the environmental impact of shipping and maritime activities, including efforts to mitigate carbon emissions, reduce pollution, and protect marine ecosystems.

21.3.2 Technological Innovation

Researchers have opportunities to explore and develop cutting-edge technologies in the maritime sector, such as autonomous vessels, renewable energy solutions, and digitalization of maritime operations, to improve efficiency and safety.

21.3.3 Safety and Security

Researchers are addressing challenges related to maritime safety and security, including the prevention of accidents at sea, cybersecurity threats, piracy, and illegal activities in maritime domains.

21.3.4 Policy and Regulation

Researchers in maritime studies face challenges related to navigating complex international regulations and policies governing maritime activities, as well as the need to develop evidence-based recommendations for policymakers to promote sustainable maritime development.

21.3.5 International Collaboration

Collaboration among researchers from different countries and regions is essential for advancing maritime research, sharing best practices, and addressing global maritime challenges that require multinational cooperation.

While researchers across various disciplines and the maritime sector face significant challenges in pushing the boundaries of knowledge, there are also numerous opportunities for collaboration, innovation, and impact in addressing critical global issues and advancing scientific understanding. By overcoming these challenges and seizing opportunities for interdisciplinary collaboration and technological innovation, researchers can make significant contributions to the advancement of knowledge and the sustainable development of society.

21.3.6 The Role of Measurement Techniques

Measurement techniques play a crucial role in maritime research, enabling scientists and researchers to collect accurate and precise data to understand various aspects of the marine environment. The advancement of measurement technology has revolutionized research in maritime by providing researchers with sophisticated tools and methods to study the oceans and seas in greater detail than ever before. Here are some critical points on the role and importance of measurement techniques in maritime research:

21.3.7 Precision and Accuracy

Measurement techniques help capture data with high precision and accuracy, crucial for making reliable observations and drawing meaningful conclusions. With precise measurements, researchers can track changes in oceanographic parameters over time and space, leading to a better understanding of marine processes.

21.3.8 Data Collection

Measurement techniques allow researchers to collect a wide range of data on various aspects of the marine environment, such as water temperature, salinity, currents, waves, nutrient levels, and biodiversity. This data is essential for studying marine ecosystems, climate patterns, and ocean dynamics.

21.3.9 Monitoring and Forecasting

Advanced measurement technologies enable continuous monitoring of marine parameters, which are vital for predicting and forecasting events like storms, tsunamis, and harmful algal blooms. Real-time data collection helps in the early detection of environmental changes and allows for timely response measures.

21.3.9.1 Sample Collection

Measurement techniques often involve collecting samples from the marine environment for further analysis. These samples can include water samples, sediment cores, biological specimens, and more. Analyzing these samples provides valuable insights into marine ecosystems' composition, structure, and health.

21.3.9.2 Advancements in Measurement Technology

Over the years, advancements in measurement technology have revolutionized maritime research by introducing innovative tools and methods. For example, the development of autonomous underwater vehicles (AUVs) and remotely operated vehicles (ROVs) has enabled researchers to explore deep-sea environments and collect data from areas that are difficult to access.

21.3.9.3 Integration of Data

Modern measurement techniques allow for integrating data from multiple sources and sensors, providing a more comprehensive understanding of the marine environment. By combining data from satellites, buoys, sensors, and other platforms, researchers can create detailed models and simulations to study complex oceanographic processes.

Indeed, measurement techniques play a vital role in maritime research by enabling the collection of precise and accurate data essential for understanding marine ecosystems, ocean dynamics, and environmental changes. The continuous advancements in measurement technology have revolutionized the field of maritime research, allowing scientists to explore and study the oceans in previously impossible ways.

21.4 The Power of Simulation Techniques

Simulation techniques are invaluable for modeling complex systems and scenarios across various domains, including maritime applications. The power of simulation lies in its ability to replicate real-world processes and behaviors in a controlled, virtual environment. Using simulations, researchers can gain insights, test hypotheses, and explore new frontiers without needing expensive real-world experiments or risking safety (Sim et.al, 2023).

In the maritime industry, simulations are crucial in various aspects such as ship design, navigation, port operations, risk assessment, and training. Here are some ways in which simulation techniques are significant in modeling complex maritime systems and scenarios:

21.4.1 Ship Design and Performance Evaluation

Simulations allow naval architects and engineers to design and optimize ships by modeling their hydrodynamics, stability, and performance under different conditions. By running simulations, researchers can test various design configurations, assess the impact of changes, and optimize ship performance before construction.

21.4.2 Navigation and Maneuvering

Simulations can replicate real-world maritime environments, including water currents, weather conditions, and traffic patterns, to help mariners train for challenging navigation scenarios. By running simulations, researchers can study the effects of different maneuvers, develop better navigation strategies, and enhance safety at sea.

21.4.3 Port Operations and Logistics

Simulations are used to model and optimize port operations, container handling, and logistics processes. Researchers can simulate the flow of ships, containers, and vehicles within a port, identify bottlenecks, optimize resource allocation, and improve overall efficiency.

21.4.4 Risk Assessment and Safety

Simulations are essential for assessing risks in maritime operations, such as collision avoidance, oil spills, and emergency response planning. By running simulations of potential scenarios, researchers can evaluate risks, develop mitigation strategies, and enhance safety protocols to prevent accidents and environmental damage.

21.4.5 Training and Education

Simulations provide a safe and cost-effective way to train maritime professionals, including ship captains, pilots, and port operators. Virtual simulations allow trainees to practice skills, experience realistic scenarios, and improve decision-making in a controlled environment without risking lives or resources.

As shown, the power of simulation techniques in modeling complex maritime systems and scenarios enables researchers to explore new frontiers, test hypotheses, and improve decision-making in a virtual environment. By leveraging simulations, researchers can gain valuable insights, optimize operations, enhance safety, and drive innovation in the maritime industry.

Some of the most cutting-edge advancements in measurement and simulation techniques shaping the future of research and development are as follows:

High-Resolution Imaging and Sensing Technologies. With the advent of advanced imaging techniques such as electron microscopy, atomic force microscopy, and hyperspectral imaging, researchers can now visualize and analyze materials at unprecedented resolutions. These capabilities are revolutionizing fields ranging from materials science to biological research, opening new frontiers for discovery.

Machine Learning and Artificial Intelligence in Simulation. The integration of machine learning and artificial intelligence algorithms into simulation models is revolutionizing prediction accuracy and efficiency. By leveraging vast amounts of data and complex algorithms, researchers can now simulate intricate systems with improved precision, enabling them to make informed decisions and optimize outcomes.

Multi-scale Modeling and Simulation. Understanding complex systems often requires multi-scale modeling approaches that bridge different spatial and temporal scales. By integrating simulation techniques at various levels of granularity, researchers can capture the emergent behavior of systems, from the molecular level to the macroscopic scale, leading to a more comprehensive understanding of phenomena.

Real-time Monitoring and Control Systems. The development of real-time monitoring and control systems enhances our ability to interact with and manipulate systems. These advancements are particularly crucial in robotics, autonomous vehicles, and industrial automation, where precise control and feedback mechanisms are essential.

Quantum Computing and Simulation. The emergence of quantum computing is poised to revolutionize simulation capabilities, particularly for complex systems beyond classical computers' reach. Quantum simulators promise to solve optimization problems, simulate quantum systems, and accelerate research in areas such as material science and drug discovery.

21.5 Synergy Between Measurement and Simulation

Combining measurement and simulation techniques in the maritime sector can significantly enhance innovation and discovery by leveraging their synergies. By integrating these approaches, researchers and industry professionals can gain a more comprehensive understanding of maritime phenomena and expedite progress in the field.

21.5.1 Comprehensive Data Collection

Measurement techniques provide real-world data on maritime parameters such as weather conditions, ship performance, and environmental factors. By combining this data with simulation techniques, researchers can create more accurate and detailed models that (Henry Sheykin, 2024) reflect the complexities of maritime operations. This comprehensive data collection enables a deeper understanding of maritime systems and processes.

21.5.2 Validation and Verification

Simulation techniques allow researchers to test hypotheses and scenarios in a controlled virtual environment. By comparing simulation results with actual measurement data, researchers can validate and verify the accuracy of their models. This iterative validation and verification process enhances simulation models' reliability and helps identify discrepancies that can lead to new insights and discoveries.

21.5.3 Predictive Analysis and Optimization

By integrating measurement and simulation techniques, researchers can develop predictive models that forecast maritime outcomes based on real-time data inputs. These predictive models enable proactive decision-making and optimization of maritime operations, improving efficiency, safety, and cost-effectiveness. Researchers can identify opportunities for innovation and process improvement by simulating various scenarios and analyzing measurement data.

21.5.4 Risk Assessment and Mitigation

Measurement techniques provide valuable data for assessing risks in maritime operations, such as collision avoidance, navigation hazards, and environmental impact. By combining this data with simulation techniques, researchers can conduct risk analyses and develop mitigation strategies to enhance maritime safety and security. The synergy between measurement and simulation techniques enables a more comprehensive risk management and emergency preparedness approach.

21.5.5 Technology Development and Design Optimisation

Measurement data can inform the design and development of maritime technologies, such as ship hulls, propulsion systems, and navigation equipment. By integrating this data into simulation models, researchers can optimize the performance of maritime systems and enhance their capabilities. This iterative design optimization process accelerates maritime sector innovation and drives the development of advanced technologies.

The synergies between measurement and simulation techniques in the maritime sector facilitate a deeper understanding of phenomena, accelerate progress, and drive innovation and discovery. By combining these approaches, researchers and industry professionals can leverage the strengths of each method to (Label, 2024) enhance decision-making, optimize operations, mitigate risks, and develop cutting-edge technologies for the maritime industry.

21.6 Impact on Global Research Landscape

Innovations in measurement and simulation techniques have significantly transformed the global research landscape in the maritime industry. These advancements have enabled researchers, engineers, and maritime professionals to understand complex phenomena better, improve efficiency, enhance safety, and drive innovation in various maritime sectors. Below are some ways in which measurement and simulation techniques have shaped the maritime industry and examples of breakthroughs and advancements:

21.6.1 Hydrodynamics Simulation

- Computational Fluid Dynamics (CFD) simulations have revolutionized the design and optimization of ships, propellers, and other maritime structures. By simulating fluid flow around ships, engineers can optimize hull shapes for improved fuel efficiency and maneuverability.
- Breakthroughs: Reduced fuel consumption and emissions through improved hull designs, optimized propeller performance, and enhanced stability of vessels in various sea conditions.

21.6.2 Structural Analysis

- Finite Element Analysis (FEA) and other structural simulation techniques help evaluate the structural integrity of maritime structures, such as ships, offshore platforms, and marine equipment.
- Breakthroughs: Enhanced safety through a better understanding of structural behavior, optimized material usage, and improved durability of maritime structures.

21.6.3 Autonomous Navigation

- Simulation technologies are crucial in testing and validating autonomous navigation systems for uncrewed surface vessels (USVs) and autonomous underwater vehicles (AUVs).
- Breakthroughs: Advancements in autonomous navigation have increased efficiency, reduced human error, and improved safety in maritime operations.

21.6.4 Environmental Impact Assessment

- Simulation tools are used to assess the environmental impact of maritime activities, such as ship emissions, underwater noise pollution, and oil spills.
- Breakthroughs: Improved environmental sustainability through better management of maritime activities and mitigation of negative impacts on marine ecosystems.

21.6.5 Weather Forecasting and Ocean Modeling

- High-fidelity simulations help predict weather patterns, ocean currents, and wave conditions, enabling better route planning and operational decision-making for maritime transportation.
- Breakthroughs: Enhanced safety at sea, optimized voyage planning, and reduced operational risks for shipping companies

21.6.6 Underwater Robotics

- Simulation techniques are instrumental in developing and testing underwater robots for various applications, including underwater exploration, maintenance of offshore structures, and marine research.
- Breakthroughs: Advanced underwater robotics have enabled the exploration of deep-sea environments, collection of valuable data, and inspection of underwater infrastructure with increased efficiency and accuracy.

These examples illustrate how innovations in measurement and simulation techniques have revolutionized the maritime industry, driving progress, efficiency, and sustainability in various aspects of maritime research and operations. The integration of advanced technologies continues to shape the maritime sector's future, offering new possibilities for exploration, optimization, and environmental stewardship.

21.7 Challenges

Adopting and implementing new measurements and simulation techniques in maritime research can be challenging due to various factors. Some of the challenges researchers may face include:

21.7.1 Cost

Acquiring new measurement equipment and implementing simulation techniques can be expensive, especially for researchers and institutions with limited budgets. The initial investment in technology and training can be a barrier to adoption.

21.7.2 *Expertise*

Utilizing advanced measurement and simulation techniques often requires specialized knowledge and skills. Researchers may need training to effectively use these tools, which can be time-consuming and require additional resources.

21.7.3 *Data Quality and Standardization*

Ensuring the quality and consistency of data obtained from new measurement techniques is crucial for research validity. Standardizing data collection methods and ensuring data accuracy can be challenging, especially when dealing with complex maritime environments.

21.7.4 *Integration with Existing Systems*

Implementing new measurement and simulation techniques may require integrating existing systems and workflows. Compatibility issues, data transfer, and interoperability challenges can arise when integrating new technologies with legacy systems.

E. Regulatory and Ethical Considerations.

Researchers in the maritime sector must comply with regulations governing data collection, storage, and sharing. Adhering to ethical standards and addressing privacy concerns when collecting and analyzing maritime data is essential but can present challenges.

21.8 Opportunities Ahead

Despite these challenges, there are significant opportunities for collaboration and interdisciplinary research in the evolving landscape of maritime research. Some potential opportunities include:

21.8.1 Interdisciplinary Collaboration

Collaborations between researchers from disciplines such as engineering, oceanography, computer science, and environmental science can lead to innovative solutions and insights in maritime research. Interdisciplinary teams can bring diverse perspectives to address complex challenges.

21.8.2 Industry Partnerships

Collaborating with industry partners can provide researchers access to real-world data, resources, and expertise. Industry partnerships can help bridge the gap between academic research and practical applications in the maritime sector.

21.8.3 International Collaboration

Maritime research often involves global challenges that require international cooperation. Collaborating with researchers from different countries can facilitate knowledge exchange, data sharing, and the development of standardized practices in maritime research.

21.8.4 Technology Transfer

Collaborating with technology providers and start-ups can facilitate the transfer of cutting-edge measurement and simulation technologies to maritime research. Researchers can benefit from industry innovations and adapt them to address specific research needs.

21.8.5 Policy and Governance Research

Interdisciplinary research that explores the policy implications of new measurement and simulation techniques in maritime can help shape regulations, standards, and governance frameworks. Researchers can contribute to shaping responsible and ethical practices in maritime research and technology adoption.

By addressing these challenges and leveraging collaboration opportunities, researchers can advance maritime research, enhance the understanding of maritime environments, and contribute to sustainable and innovative solutions for the maritime industry.

21.9 Concluding Remarks

In conclusion, we are encouraged to embrace innovation, collaboration, and creativity in work. Our contributions are essential for advancing the understanding of the maritime world and addressing its challenges. By pushing the boundaries of knowledge and exploring new research frontiers, we can significantly impact the field.

Innovation is critical to developing new technologies, methodologies, and solutions to improve maritime safety, efficiency, and sustainability. Embracing creativity in work can lead to novel ideas and approaches that have the potential to revolutionize the way we interact with the oceans.

Collaboration with experts from different disciplines and backgrounds can bring fresh perspectives and insights to our research. By working together with others, we can leverage diverse expertise and resources to tackle complex problems and make breakthrough discoveries.

Let us stay curious, keep an open mind, and never be afraid to think outside the box. Our passion for exploration and discovery is what drives progress in maritime research. By continuously challenging ourselves and each other, we can unlock new opportunities and contribute to shaping a better future for the maritime industry and the world.

Remember, the huge sea is full of mysteries, waiting to be unraveled. Let our imagination soar, and our research pave the way for a brighter tomorrow.

Keep pushing the boundaries, keep exploring new frontiers, and keep inspiring others with our work. The future of maritime research is in our hands, and we will rise to the challenge.

Indeed, the advancements in measurement and simulation techniques we will witness today will empower researchers to push the boundaries of knowledge and drive innovation across disciplines. As we continue to explore new frontiers and tackle grand challenges, the integration of these advanced tools will be instrumental in shaping the future of research in maritime.

The Institute of Electrical and Electronics Engineers (IEEE) conference is an excellent venue for the sharing of ideas and work together on various exciting opportunities and enlightening discussions for research endeavors.

References

1. Why are online simulations such a powerful educational tool? SimTutor Limited. https://www.simtutor.com/blog/posts/2023/may/why-are-online-simulations-such-a-powerful-educational-tool/. Accessed 15 Aug 2024
2. Improve financial modeling outputs with these proven optimization strategies. https://finmodelslab.com/blogs/blog/optimizing-financial-modeling-outputs. Accessed 15 Aug 2024
3. Data labeling: techniques, best practices and future trends. https://urdustem.com/exploring-data-labeling-techniques-and-best-practices/. Accessed 15 Aug 2024

Chapter 22
The Concept of Green Management for Improving the Environmental Performance of Traditional Energy Facilities

Anzhelika Karaeva, Florentina-Aurelia Vasilica, Vincenzo Torretta, and Elena Cristina Rada

Abstract Improving the efficiency of environmental protection activities at traditional energy facilities is one of the key directions of eco-modernization of the industry in the next decade. The introduction of the best available environmental technologies, in addition to the need to comply with the SDGs, the transition to a circular economy and the general tightening of environmental policy with the slow renewal of outdated energy infrastructure poses new management challenges that hinder the improvement of environmental performance in the energy industry. It is necessary to improve current approaches to managing traditional energy facilities, develop new management concepts and introduce new tools for organizing internal business processes. Modern management should be aimed not only at achieving certain socio-economic results, but also at greening current activities. The purpose this study is to develop a concept of green management for traditional energy facilities. A list of criteria has been formed and a concept of green management with appropriate management tools has been proposed.

Keywords Traditional energy · Green management · Environmental protection · Energy sector · Circular economy

A. Karaeva (✉) · V. Torretta · E. C. Rada
Department of Theoretical and Applied Sciences, University of Insubria, Varese, Italy
e-mail: akaraeva@uninsubria.it

F.-A. Vasilica
National University of Science and Technology POLITEHNICA, Doctoral School of Engineering Power, Bucharest, Romania

R. Tamas and C. A. Heiman (eds.), *Selected Papers from AToMS 2024*,
Lecture Notes on Multidisciplinary Industrial Engineering,
https://doi.org/10.1007/978-3-032-02452-7_22

22.1 Introduction

The energy sector is a critical element of the normal functioning of the economy and society and has a role to play in ensuring economic growth. However, management in energy differs from management in other industries due to several sectoral features such as [1–4]:

- The high degree of state regulation;
- Electricity and heat generation must be consistent with constantly changing demand;
- The use of significant amount of fossil fuels (relevant for traditional energy facilities);
- Significant environmental impact;
- The need to comply with the international environmental agenda.

Current challenges and branch features require a considerable review of existing management approaches. The effective functioning of national energy systems and rapid eco-modernization of the global energy sector is impossible without developing new adapted managerial approaches and tools at traditional energy facilities. All those must include the Circular Economy (CE) concepts, Ecodesign for Sustainable Products Regulation (ESPR) law indication and possible also to comply with FIT for 55 requirements also considering the public attitude to all [5–7].

Green management is a management concept in which reducing the negative impact of an enterprise on the environment while increasing its commercial and production efficiency comes to the fore [8, 9]. Green management can also be described as the key component of sustainable management.

The terms "green management" and "sustainable management" are increasingly found in scientific and professional literature in the field of management every year. However, despite the growing interest in sustainable development and greening of industry, there is currently no unified interpretation of the concept of "sustainable management", which includes the concept of green management. Many scientists consider sustainable management as a set of management tools such as corporate governance, environmental management and corporate social responsibility, which corresponds to the economic, environmental and social aspects of the concept of sustainable development [10–14]. The paper [14] provides the opinion that sustainable management is management aimed primarily at reducing the negative impact on the environment. However, according to the concept, sustainable development can be achieved only in the balance of all three components: achieving environmental sustainability alone cannot guarantee the sustainability of an enterprise in the long term. Some authors consider sustainable management through the prism of managing various enterprise resources (economic, labor, natural, etc.) [15]. This approach has a number of practical advantages such as the possibility of tactical planning of the enterprise's activities in accordance with the principles of sustainable development and the implementation of activities taking into account the resources available to the enterprise.

Thus, green management is an integral part of sustainable management. Environmental problems of energy, the climate agenda and the increasing demand for energy require the development of industry-specific management approaches, where the environmental aspects of the functioning of energy facilities will come to the fore. At the same time, green management should take into account the importance of achieving positive financial and economic results in order to maintain the economic stability of an energy enterprise in the long term.

This study aims to design the basis for the concept of green management at energy enterprises that would consider economic and technical features and environmental challenges that energy sector faces.

22.2 Materials and Methods

The present study included the following stages:

I. Review of scientific literature on the topic of green and sustainable management; definition of the role of green management in the sustainable management system.
II. Analysis of the specifics of traditional energy and its impact on the environment.
III. Formation of the initial concept of green management in traditional energy: definition of the goals and objectives of green management, formation of a list of key components of green management and appropriate tools.

Authors used the following research methods: scientific literature review, analysis of environmental reports published by energy companies, general scientific and systematic approaches.

22.3 Results and Discussion

The main purpose of green management in the energy sector is reduction of a huge environmental impact of traditional energy enterprises while simultaneously meeting the growing demand for energy, provided stable access to heat and electricity and maintaining an acceptable price level.

The objectives of green management in the energy sector include:

- Reducing and achieving the minimum possible negative impact of an energy facility on all natural environments;
- Improving the resource efficiency of energy production;
- Development of short-term and long-term environmental policy, drawing up plans and a list of measures to improve the effectiveness of environmental protection activities;

- Regular environmental impact assessment of the energy facility's activities (for example, once a year or quarterly, depending on the size of the energy enterprise and the type of fuel used);
- Regular analysis of the compliance of current activities with relevant environmental standards and norms;
- Annual compilation and publication of environmental reports in the public domain;
- Continuous training of employees working in the production and environmental departments.

According to the results of the study, it was concluded that the concept of green management at energy facilities includes four groups of components:

- Rational use of natural resources for energy production (natural energy carrier (gas, oil or cool), water resources, land and soil resources);
- Greening of production processes;
- Greening of organizational processes;
- Continuous innovative activity in the field of environmental protection.

Table 22.1 shows the key elements of green management of energy enterprises.

The key element of green management is the EMS, which enables to consider the priorities of the company's environmental activities and plan future environmental activities for the short and long term. EMS is mostly used in large enterprises, but their simplified introduction is possible in small and medium-sized energy enterprises. For example, the ISO 14000 series of standards "Environmental Management" [24, 25] can become the basis for the installation of EMS and, consequently, the basis of the concept of green management: the standards contain the basic principles, criteria and a flexible approach to the step-by-step implementation of the management system. In addition, EMS involves the restructuring and greening of the main business processes and professional development of employees in the field of ecology and environmental protection.

The concept of sustainable management implies a type of management that achieves a balance between environmental, social and economic sustainability. Without achieving a balance between environmental impact, financial performance and social well-being of employees, it is impossible to achieve the sustainability of the enterprise. However, if the tools for achieving economic and social sustainability in different sectors of the economy are relatively similar (effective capital management of an enterprise, effective investment activities, government support and subsidies, corporate social responsibility (CSR), improving the quality of working conditions, etc.), then in order to achieve environmental sustainability it is necessary to develop new approaches and tools that take into account the industry specifics [22–24, 26, 27]. Green management can be indirectly identified with the concept of environmental sustainability of energy enterprises. Based on this, the role of green management and its relationship with other elements of sustainable management is shown in Fig. 22.1.

Table 22.1 Key elements of green management concept in energy sector [16–23]

Component of green management	Key elements/tools
Rational use of natural resources for energy production	• Implementation of the circular economy (CE) principles; • Availability of an environmental management system (EMS); • Implementation of resource- and energy-saving technologies; • Timely maintenance and replacement of production equipment with a high degree of depreciation; • Ensuring stable access to natural resources
Greening of production processes	• Using the best available technologies (BAT) in production processes and environmental protection activities; • Implementation of resource- and energy-saving technologies; • Availability of an EMS; • Control over the depreciation of fixed assets and their timely replacement; • Timely maintenance and replacement of production equipment with a high degree of depreciation; • Environmental risk management and development of measures to minimize negative environmental consequences in case of emergency situations
Greening of organizational processes	• Availability of an EMS; • Implementation of the principles of the circular economy; • Conducting environmental audits and monitoring; • Development of the company's environmental policy; • Staff training; • Continuous analysis of changes in environmental legislation
Continuous innovative activity in the field of environmental protection	• Continuous improvement of environmental efficiency through the development and implementation of new technologies, principles and approaches to management; • Attracting third-party investments and (or) government support for periodic renewal and expansion of production capacities

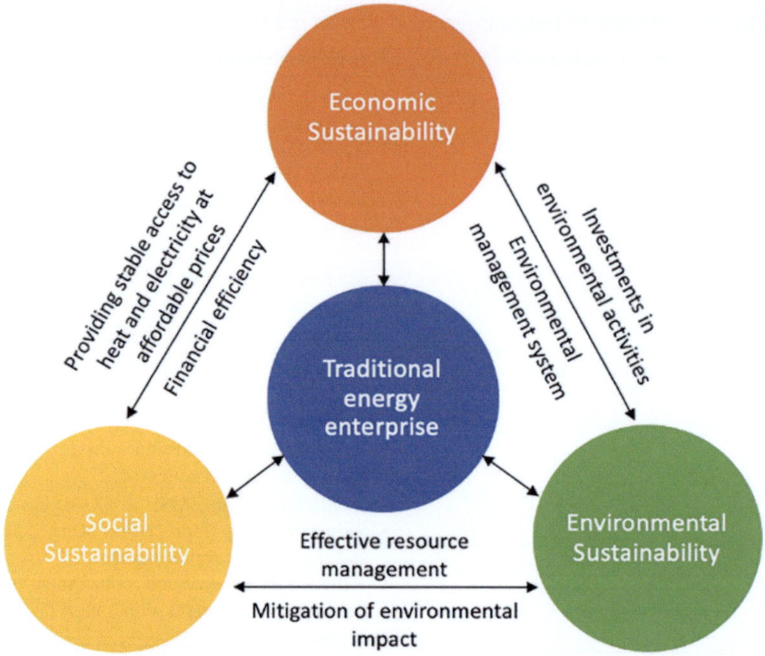

Fig. 22.1 Interrelation of sustsainable management componnets (made by the authors)

The implementation of environmental protection measures and the eco-modernization of traditional energy facilities also requires significant financial, material and labor costs. Due to the central role of the energy industry in ensuring the normal functioning of the economy, energy companies receive financial support from the state for the implementation of various energy projects and environmental protection measures.

The introduction of the green management concept and appropriate tools contribute to reducing anthropogenic pressure in the near future and, therefore, can lead to significant economic benefits: reducing the amount of environmental payments, increasing the competitiveness and investment attractiveness of energy facilities, accelerating the introduction of the principles of circular economy into the main production and business processes, etc.

22.4 Conclusions

Green management can become one of the drivers of greening the energy sector in the foreseeable future. However, in order to achieve significant results in the field of ecology and environment, it is necessary to develop adapted management approaches and tools that will take into account industry and regional specifics. Within the framework of this work, the main goals, objectives, components and relevant tools have been formed, which form the initial concept of green management for traditional energy enterprises.

References

1. Karaeva, A., Ionescu, G., Cioca, L.I., Tolkou, A., Kyzas, G.: Environmental sustainability for traditional energy small and medium enterprises. Environ. Sci. Pollut. Res. **30**, 47822–47831 (2023). https://doi.org/10.1007/s11356-023-25718-x
2. Cariola, A., Fasano, F., Rocca, M., Skatova, E.: Environmental sustainability policies and the value of debt in EU SMEs: Empirical evidence from the energy sector. J. Clean. Prod. **275**, 123133 (2020). https://doi.org/10.1016/j.jclepro.2020.123133
3. Karaeva, A., Magaril, E., Al-Kayiem, H., Torretta, T., Rada, E.C.: Approaches to the assessment of ecological and economic efficiency of investment projects: brief review and recommendations for improvements. WIT Transact Ecol Environ. **253**, 515–525 (2021). https://doi.org/10.2495/SC210421
4. Karaeva, A., Magaril, E.: Considering waste generation in the energy sector during the transition to a circular economy. Recycling **8**(2), 42 (2023). https://doi.org/10.3390/recycling8020042
5. Karaeva, A., Magaril, E., Torretta, V., Viotti, P., Rada, E.C.: Public attitude towards nuclear and renewable energy as a factor of their development in a circular economy frame: two case studies. Sustainability **14**(3), 1283 (2022). https://doi.org/10.1016/j.egyr.2021.08.101
6. Vezzoli, C., Macrì, L.: The design of sustainable product-service systems to foster circular economy for all. Greening Ind Networks Stud **12**, 39–64 (2024). https://doi.org/10.1007/978-3-031-55036-2_3
7. Ullman, A.N., Kittner, N.: Are global efforts coordinated for a Just Transition? a review of civil society, financial, government, and academic just transition frameworks. Energy Res. Soc. Sci. **108**, 103371 (2024). https://doi.org/10.1016/j.erss.2023.103371
8. Raut, R.D., Luthra, S., Narkhede, B.E., Mangla, S.K., Gardas, B.B., Priyadarshinee, P.: Examining the performance oriented indicators for implementing green management practices in the Indian agro sector. J. Clean. Prod. **215**, 926–943 (2019). https://doi.org/10.1016/j.jclepro.2019.01.139
9. Ma, Y., Xue, L.: Organizational legitimacy and firm nonmarket strategies as catalysts of alliance green management. J. Clean. Prod. **447**, 141315 (2024). https://doi.org/10.1016/j.jclepro.2024.141315
10. Johnson, M.P., Schaltegger, S.: Two decades of sustainability management tools for SMEs: How Far Have We Come? J. Small Bus. Manag. **54**, 481–505 (2014). https://doi.org/10.1111/jsbm.12154
11. Rebelo, M., Santos, G., Silva, R.: Integration of individualized management systems (MSs) as an aggregating factor of sustainable value for organizations: an overview through a review of the literature. J Mod Acc Audit **10**, 356–383 (2014)
12. Mežinska, I., Lapiņa, I., Mazais, J.: Integrated management systems towards sustainable and socially responsible organization. Total Qual. Manag. Bus. Excell. **26**, 469–481 (2015). https://doi.org/10.1080/14783363.2013.835899
13. Vnoučková, L., Hyršlová, J., Tomšík, P.: How does implementation of environmental management system contribute to corporate sustainability management. Acta Univ Agric Silvic Mendel Brun **62**, 1499–1508 (2014). https://doi.org/10.11118/actaun201462061499
14. Kosacka, M., Werner-Lewandowska, K.: Methodology for sustainable resource management measurement at a company level—case study. Procedia Manufacturing **11**, 725–733 (2017). https://doi.org/10.1016/j.promfg.2017.07.173

15. Barbosa, M., Castañeda-Ayarza, J.A., Ferreira, D.H.L.: Sustainable strategic management (GES): sustainability in small business. J. Clean. Prod. **258**, 120880 (2020). https://doi.org/10.1016/j.jclepro.2020.120880
16. Kashi, B.: Risk management and the stated investment costs by independent power producers. Energy Econ **49**, 660–668 (2015). https://doi.org/10.1016/j.eneco.2015.02.022
17. Klug, T.W., Beyene, A.D., Meles, T.H., Toman, M.A., Hassen, S., Hou, M., Klooss, B., Mekonnen, A., Jeuland, M.: A review of impacts of electricity tariff reform in Africa. Energ Policy **170**, 113226 (2022). https://doi.org/10.1016/j.enpol.2022.113226
18. Fuentes, S., Villafafila-Robles, R., Olivella-Rosell, P., Rull-Duran, J., Galceran-Arellano, S.: Transition to a greener Power Sector: Four different scopes on energy security. Renew Energ Focus **33**, 23–36 (2020). https://doi.org/10.1016/j.ref.2020.03.001
19. Liu, Y., Dong, K., Dong, X., Taghizadeh-Hesary, F.: Towards a sustainable electricity industry in China: An appraisal of the efficacy of environmental policies. Util Policy **86**, 101700 (2024). https://doi.org/10.1016/j.jup.2023.101700
20. Al-Shetwi, A.Q.: Sustainable development of renewable energy integrated power sector: Trends, environmental impacts, and recent challenges. Sci. Total. Environ. **822**, 153645 (2022). https://doi.org/10.1016/j.scitotenv.2022.153645
21. Su, Z., Yang, L.: Energy management and life cycle assessment of efficient and flexible trigeneration system for coal-fired power plants. Appl. Therm. Eng. **217**, 119178 (2022). https://doi.org/10.1016/j.applthermaleng.2022.119178
22. Li, S.: Natural extreme events, government subsidies and corporate environment responsibility: Evidence from China's energy sector. Energy Econ **114**, 106278 (2022). https://doi.org/10.1016/j.eneco.2022.106278
23. Karaeva, A., Copelli, S., Ivascu, L., Lakatos, S.E., Romagnoli, F., Torres, T.A.: Introducing principles of circular economy into traditional energy sector: barriers and prospectives. In: 11th international conference on energy and environment, CIEM. https://doi.org/10.1109/CIEM58573.2023.10349713
24. Karaeva, A., Magaril, E., Torretta, V., Ragazzi, M., Rada, E.C.: Green energy development in an industrial region: a case-study of Sverdlovsk region. Energy Rep. **7**, 137–148 (2021). https://doi.org/10.1016/j.egyr.2021.08.101
25. ISO 14000 Family Environmental Management. https://www.iso.org/standards/popular/iso-14000-family
26. Karaeva, A., Tolkou, A., Cioca, L.I., Lakatos, E.S.: Family ISO 14000 standards as a tool of achieving environmental sustainability of enterprises. IOP Conf. Ser.: Earth Environ. Sci. **126**(1), 012036 (2023). https://doi.org/10.1088/1755-1315/1126/1/012036
27. Brookes, N.J., Locatelli, G.: Power plants as megaprojects: Using empirics to shape policy, planning, and construction management. Util Policy **36**, 57–66 (2015). https://doi.org/10.1016/j.jup.2015.09.005

Chapter 23
Enhancing Operational Efficiency and Environmental Sustainability by Improving the Maintenance Process in an Industrial Organization

Ioan Cristian Nicolae, Mihai Victor Zerbes, Liliana Georgeta Popescu, Ramona Giurea, Anzhelika Karaeva, Fabio Conti, Vincenzo Torretta, and Elena Cristina Rada

Abstract In response to the imperative of enhancing operational efficiency and environmental sustainability, this study employs the principles of Six Sigma to improve maintenance activities for weaving machines in an airbag manufacturing facility. The project's objective is to refine maintenance processes, specifically targeting the elimination of rework resulting from faulty maintenance practices. Following the

I. C. Nicolae
Asociația Conexiuni, Sibiu, Romania
e-mail: ioancristiannicolae@yahoo.com

M. V. Zerbes · L. G. Popescu · R. Giurea
Department of Industrial Engineering and Management, Lucian Blaga University of Sibiu, Sibiu, Romania
e-mail: mihai.zerbes@ulbsibiu.ro

L. G. Popescu
e-mail: liliana.popescus@ulbsibiu.ro

R. Giurea
e-mail: ramona.giurea@ulbsibiu.ro

A. Karaeva (✉) · F. Conti · V. Torretta · E. C. Rada
Department of Theoretical and Applied Sciences, University of Insubria, Varese, Italia
e-mail: akaraeva@uninsubria.it

F. Conti
e-mail: fabio.conti@uninsubria.it

V. Torretta
e-mail: vincenzo.torretta@uninsubria.it

E. C. Rada
e-mail: elena.rada@uninsubria.it

R. Tamas and C. A. Heiman (eds.), *Selected Papers from AToMS 2024*,
Lecture Notes on Multidisciplinary Industrial Engineering,
https://doi.org/10.1007/978-3-032-02452-7_23

DMAIC framework, the research identified the root causes of maintenance inefficiencies through the voice of the customer, initiating a structured approach towards optimization. The results demonstrate significant improvements across various parameters. Notably, downtime decreased, productivity increased, and the quality of the finished product improved. Furthermore, the project facilitated enhanced training for mechanics, fostering a culture of continuous improvement within the organization. Importantly, these enhancements have profound positive effects on the environment. By optimizing energy usage and improving mechanical systems, the project reduces carbon emissions and resource consumption, contributing to a more sustainable manufacturing process. However, it's essential to note that the results of the implementation of this project are yet to be measured, with ongoing evaluation essential to determine its full impact on environmental and operational metrics.

Keywords Environmental sustainability · Process · Maintenance · Organization · Six Sigma · DMAIC

23.1 Introduction

The problem from which the study presented in this article starts, refers to the implementation of multiple unplanned maintenances (unplanned down-time). Machine failure occurs immediately after maintenance, when the steps to be followed in one of the four maintenances are not met. These problems occur in the weaving section, and the responsible persons are the mechanics on the section. Due to problems with the weaving machines, production will slow down. The quantity and quality of materials used will decrease. An important factor to consider is productivity (quantity produced/person). If there is poor communication between departments, conflicts will arise between employees. The methodology for ameliorating this problem was DMAIC [1].

23.2 Methodology

Effective process improvement methodologies are critical for maintaining and enhancing operational efficiency, particularly in complex manufacturing environments. Recognizing the need for a structured approach to reduce unplanned downtimes in their weaving section, a leading automotive safety systems manufacturer adopted the Six Sigma DMAIC (Define, Measure, Analyze, Improve, Control) methodology, which is renowned for its systematic and data-driven approach to process improvement [2].

The Six Sigma DMAIC methodology provides a robust framework for identifying and eliminating defects in any process, ensuring consistent quality and performance. By focusing on critical-to-quality (CTQ) characteristics and utilizing a variety of statistical tools, DMAIC helps organizations achieve significant improvements in their processes. This methodology is particularly effective in manufacturing settings, where even minor inefficiencies can lead to substantial costs and downtime [3].

The implementation of the DMAIC methodology was driven by a cross-functional team comprising a project coordinator, production engineers, quality engineers, and maintenance personnel. The primary goal was to reduce unplanned maintenance activities to 5% of total maintenance within six months. This ambitious target required a comprehensive understanding of the existing process, identification of key problem areas, and development of targeted solutions [4].

The following sections detail the step-by-step approach taken, beginning with the Define phase, where the project's scope and objectives were established, followed by the Measure phase, which involved data collection and baseline performance assessment. The Analyze phase focused on identifying root causes of inefficiencies, leading to the Improve phase, where solutions were developed and tested. Finally, the Control phase ensured that the improvements were sustained and continuously monitored. Through this structured approach, the organization aimed not only to enhance operational efficiency but also to foster a culture of continuous improvement and excellence.

Define.

The Define phase is crucial as it sets the foundation for the Six Sigma project. The primary objective was to minimize unplanned downtimes in the weaving section of the plant. A cross-functional team was assembled, including a project coordinator, production engineer, quality engineer, and the shift leader of the preventive maintenance team. The project aimed to reduce unplanned maintenance to just 5% of total maintenance activities within six months of implementation. The key deliverables for this phase included a detailed project charter, a high-level process map, and a clear definition of customer requirements.

To define the problem correctly a problem assessment matrix has been created. The main problem that stood out, after calculating the weighted average for five different issues, was that 30% of the maintenance performed on the weaving machines were unplanned.

A vital aspect of the Define phase is stakeholder engagement. The team conducted several meetings with key stakeholders to understand their expectations and gather input. This ensured that the project goals were aligned with the overall business objectives. Additionally, a SIPOC (Suppliers, Inputs, Process, Outputs, Customers) diagram was created to provide a high-level view of the process and identify the critical elements that impact performance. This structured approach helped in setting clear and achievable goals for the project.

Measure.

The Measure phase involved gathering data to establish a baseline performance and identify the scope of the problem. For the weaving machines, four preventive maintenance activities were scheduled to ensure optimal operation. Data was collected on unplanned interventions, which were then categorized and analyzed to understand the frequency and impact of these downtimes. The team employed various measurement tools and techniques, including process mapping and statistical analysis, to quantify the current state of machine performance and identify critical areas for improvement.

One of the significant tasks in this phase was the development of a detailed data collection plan. The team ensured that the data collected was accurate, relevant, and sufficient to provide insights into the problem. Tools such as control charts and capability analysis were used to understand the variability in the process and establish performance benchmarks. By meticulously measuring and documenting the existing conditions, the team could pinpoint where deviations occurred and the extent of their impact on the overall process.

The data collected and analyzed were over a period of one year. The analysis of the data brought to light, that after the preventive maintenance 1 and 3 the most unplanned interventions were required. This is how situations arise in which maintenance is not carried out or is carried out superficially, due to the overlapping of maintenance that takes place simultaneously. In order to understand how this process works, a flow chart has also been done.

Analyze.

In the Analyze phase, the team sought to identify the root causes of unplanned downtimes. They used tools such as cause-and-effect diagrams, Pareto charts, and failure mode and effects analysis (FMEA). The primary findings indicated that maintenance workers often missed crucial steps due to the complexity and clutter of the maintenance checklists. Additionally, improper lubrication was identified as a significant factor leading to premature component failures. By streamlining the maintenance procedures and providing clear, visual instructions, the team could pinpoint the exact reasons for machine failures and areas requiring immediate attention.

The Analyze phase also involved conducting detailed process analysis and hypothesis testing. The team applied statistical tools to validate their hypotheses and ensure that the identified causes were indeed the primary contributors to the problem. Regression analysis and design of experiments (DOE) were among the techniques used to explore relationships between variables and determine the most significant factors affecting performance. This rigorous analytical approach ensured that the solutions developed were based on solid evidence and targeted the real issues.

A Pareto chart was made by estimating the weight of causes that led to machine failure. In order to make the Pareto chart, it was necessary to rank the causes that led to the highlighting of the priority problems, according to the weight of each one.

From the Pareto chart the first 4 causes generate 81% of all defects are:

- Cluttered maintenance sheet;
- Imbalance of the activities be carried out in the 4 maintenances;
- Insufficient staff training.
- Uneven scheduling of the weekly maintenances

Improve.

The Improve phase focused on implementing solutions to address the root causes identified in the Analyze phase. The team revised the maintenance checklists, eliminating redundant information and adding visual aids to enhance clarity and compliance. They also established a dedicated lubrication team responsible for ensuring that all components were adequately lubricated. Training sessions were conducted to upskill maintenance workers on the new procedures, and a pilot test was carried out to validate the effectiveness of these improvements before full-scale implementation.

To ensure the solutions were effective and sustainable, the team adopted a systematic approach to implementation. Pilot testing allowed them to refine the improvements and address any unforeseen issues before rolling them out across the entire operation. Additionally, mistake-proofing techniques, such as poka-yoke, were applied to prevent errors and ensure that the new procedures were foolproof. Continuous monitoring during this phase helped in making necessary adjustments and optimizing the process further.

23.2.1 Control

In the Control phase, measures were put in place to sustain the improvements achieved. A control plan was developed, including regular audits and performance reviews to monitor the adherence to the new maintenance procedures. Key performance indicators (KPIs) were established to track the reduction in unplanned downtimes, and a feedback loop was created to continually assess and refine the maintenance processes. By maintaining rigorous oversight and continuous improvement practices, the organization ensured that the gains from the Six Sigma project were preserved and built upon over time.

The Control phase emphasized the importance of standardization and documentation. Standard operating procedures (SOPs) were updated to reflect the new processes, and comprehensive training materials were developed to ensure consistent implementation. The team also leveraged technology, such as maintenance management software, to automate tracking and reporting, making it easier to maintain control over the process. Regular review meetings and performance dashboards provided visibility into the ongoing performance, enabling proactive management and continuous improvement.

By following the DMAIC methodology, the organization effectively reduced unplanned downtimes, improved operational efficiency, and enhanced overall productivity. This structured approach not only helped in significantly reducing unplanned downtimes but also fostered a culture of continuous improvement and operational excellence within the organization (Table 23.1).

Table 23.1 Identified solutions to the problems

Cause	Solution 1	Solution 2	Solution 3
Overcrowded maintenance sheet	Creating a new maintenance sheet	Adding photographs to replace written explanations	Developing a maintenance check-sheet that includes only the most frequent activities encountered in unplanned interventions
Imbalance of activities that need to be performed in the 4 maintenances	Adding additional maintenance to balance activities	Establishing a special team of workers to handle component lubrication	Removing low-priority activities
Insufficient training of personnel	Appointing a person to exclusively handle the training of new employees	Establishing a special department responsible for employees with less than one year of experience	Implementing a more rigorous recruitment process

23.3 Results and Discussion

Within our problem by applying this tool we reached the following results:

- The main cause of the problems was identified, which was the fact that the maintenance sheet is cluttered and the information it contains is difficult to follow. Due to the congestion on the maintenance sheet, the workers omit to perform all the steps present on the list. To optimize the maintenance sheet, the following changes were made: redundant information was removed and photos were added to explicitly present the process of solving the problem;
- It was observed after the analysis that the main cause of premature failure of the components on the grinding machines is the lack of lubrication or incorrect lubrication. Thus, it was decided to establish a team of workers, whose main purpose will be to verify and implement the best possible greasing of the components. In order not to lose sight of the most important steps in the maintenance process, a check-sheet was created that will include the most important actions that must be taken into account;
- Finally, the implementation plan of the improvement solutions identified as the most effective was developed, where the resources, the responsible persons and the deadlines were identified;
- In addition to these results, after the implementation of this plan, the improvement achieved by the decrease in the number of unplanned interventions, which was reduced by approximately 30% and implicitly in the related costs, can best be observed.

23.4 Conclusions

1. Enhanced Operational Efficiency: The implementation of the Six Sigma DMAIC methodology significantly improved operational efficiency by reducing unplanned downtimes in the weaving section. This reduction not only increased productivity but also optimized resource utilization. The methodology's structured approach allowed for the identification and elimination of inefficiencies, leading to more streamlined operations. By systematically addressing the root causes of downtime, the organization was able to ensure that machines operated at peak efficiency, minimizing disruptions and maximizing output.
2. Effective Problem Solving: Utilizing tools such as cause-and-effect diagrams, Pareto charts, and failure mode and effects analysis (FMEA), the project effectively identified and addressed the root causes of maintenance issues. This systematic problem-solving approach is a core benefit of the Six Sigma methodology, ensuring that solutions are based on data and thorough analysis. By understanding the underlying factors contributing to downtimes, the team was able to implement targeted interventions that addressed the specific issues, resulting in more sustainable and long-term improvements.
3. Stakeholder Engagement: Engaging stakeholders throughout the project ensured that the objectives were aligned with organizational goals. This alignment facilitated smoother implementation and higher acceptance of new processes. Stakeholder involvement was critical in gathering insights and feedback, which helped tailor the project to meet actual needs and expectations. By fostering a collaborative environment, the project team was able to build support and commitment from all levels of the organization, enhancing the overall success of the initiative.
4. Sustainability Through Reduced Waste: By minimizing unplanned downtimes and optimizing maintenance activities, the project contributed to environmental sustainability. Efficient operations lead to reduced waste and energy consumption, which are critical components of sustainable manufacturing practices. The reduction in unplanned interventions meant that resources were used more efficiently, and the overall environmental footprint of the operations was decreased. This aligns with broader sustainability goals, emphasizing the importance of responsible resource management in manufacturing.
5. Improved Maintenance Procedures: Revising and streamlining maintenance checklists and establishing a dedicated lubrication team significantly improved maintenance procedures. This not only enhanced machine reliability but also ensured that maintenance practices were more environmentally sustainable by reducing the likelihood of excessive lubrication and oil waste. By implementing clear and concise maintenance protocols, the team reduced the complexity and variability of maintenance tasks, leading to more consistent and effective maintenance outcomes.

6. Training and Upskilling: Conducting training sessions to upskill maintenance workers on new procedures ensured consistent and effective implementation of improvements. This focus on continuous learning is essential for maintaining high standards of process efficiency and sustainability within the organization. By investing in employee development, the organization ensured that workers were equipped with the necessary skills and knowledge to perform their tasks effectively, contributing to the overall success and sustainability of the improvements.
7. Standardization and Documentation: Updating standard operating procedures (SOPs) and developing comprehensive training materials ensured that improvements were well-documented and standardized across the organization. This standardization is crucial for maintaining control over processes and sustaining improvements in the long term. Clear and detailed documentation provided a reference for best practices, ensuring that all employees followed the same procedures, which helped maintain consistency and quality in operations.
8. Data-Driven Decision Making: The rigorous use of data collection, statistical analysis, and hypothesis testing during the project exemplifies the importance of data-driven decision-making in Six Sigma. This approach ensures that changes are based on solid evidence, leading to more reliable and sustainable process improvements. By relying on quantitative data to inform decisions, the project team was able to implement changes that were not only effective but also scientifically validated, enhancing the credibility and impact of the interventions.
9. Environmental Benefits of Process Optimization: The project's focus on reducing unplanned downtimes and optimizing maintenance not only improved operational efficiency but also had positive environmental impacts. Efficient processes reduce the environmental footprint by lowering energy consumption and minimizing waste, contributing to the organization's sustainability goals. By optimizing the use of resources and reducing unnecessary interventions, the project supported the organization's commitment to environmental stewardship and sustainable manufacturing practices.
10. Culture of Continuous Improvement: The successful implementation of the DMAIC methodology fostered a culture of continuous improvement within the organization. This culture encourages ongoing evaluation and enhancement of processes, ensuring that the benefits of the project are sustained and built upon, supporting both operational excellence and environmental sustainability. By embedding the principles of Six Sigma into the organizational culture, the company created a proactive environment where continuous improvement became a standard practice, driving long-term success and sustainability.

By integrating the Six Sigma DMAIC approach, the organization not only achieved its goal of reducing unplanned downtimes but also enhanced overall process efficiency and sustainability. The project highlights the importance of systematic, data-driven methodologies in achieving both operational and environmental goals in modern manufacturing. Through rigorous analysis, targeted interventions, and a commitment to continuous improvement, the organization demonstrated how structured methodologies can drive significant and sustainable improvements in manufacturing processes.

References

1. Zerbes, M.V., Popescu, L.G., Olteanu, M.: Continuous improvement of the quality management system within an automotion company in Romania. In: CoSME'20–Proceedings published in IOP conference series: materials science and engineering. ISSN 1009012061, 1-12. https://doi.org/10.1088/1757-899X/1009/1/012061
2. Zerbes M. V.: Îmbunătățirea proceselor organizaționale—Modelul DMAIC, Sibiu: Editura Techno Media S.R.L. din Sibiu. ISBN 978-606-616-371-2 (2019)
3. Kifor, C.V., Oprean, C.: Ingineria Calităţii. Îmbunătăţirea şase sigma, Sibiu, Editura Universităţii Lucian Blaga din Sibiu (2006). ISBN (10) 973-739-035-0, ISBN (13) 978-973-739-035-6
4. Joyson Safety Systems Sibiu SRL.: Internal technical documentation (2022)

MIX
Papier aus verantwortungsvollen Quellen
Paper from responsible sources
FSC® C105338

If you have any concerns about our products,
you can contact us on
ProductSafety@springernature.com

In case Publisher is established outside the EU,
the EU authorized representative is:
Springer Nature Customer Service Center GmbH
Europaplatz 3, 69115 Heidelberg, Germany

Printed by Libri Plureos GmbH
in Hamburg, Germany